LECTURES ON ION-ATOM COLLISIONS
FROM NONRELATIVISTIC TO RELATIVISTIC VELOCITIES

LECTURES ON
ION-ATOM COLLISIONS
FROM NONRELATIVISTIC
TO RELATIVISTIC
VELOCITIES

JÖRG EICHLER
Bereich Theoretische Physik, Hahn-Meitner-Institut Berlin,
14109 Berlin, Germany
and
Fachbereich Physik, Freie Universität Berlin,
14195 Berlin, Germany

2005
ELSEVIER
Amsterdam – Boston – Heidelberg – London – New York – Oxford
Paris – San Diego – San Francisco – Singapore – Sydney – Tokyo

ELSEVIER B.V.	ELSEVIER Inc.	ELSEVIER Ltd	ELSEVIER Ltd
Radarweg 29	525 B Street, Suite 1900	The Boulevard, Langford Lane	84 Theobalds Road
P.O. Box 211, 1000 AE Amsterdam	San Diego, CA 92101-4495	Kidlington, Oxford OX5 1GB	London WC1X 8RR
The Netherlands	USA	UK	UK

First edition 2005

Library of Congress Cataloging in Publication Data
A catalog record is available from the Library of Congress.

British Library Cataloguing in Publication Data
A catalogue record is available from the British Library.

ISBN 13: 978-0-444-52047-0

Transferred to Digital Printing 2007

To
Midori

Preface

The present introduction to ion-atom collisions evolved from lectures addressed to graduate students and professional scientists who were interested in atomic collisions at nonrelativistic and relativistic energies. These collisions bring some unique opportunities to study atomic structure and reaction mechanisms in experiment and theory, especially for ions of the high atomic charge numbers Z provided by modern accelerators.

The field is so vast that an adequate presentation is not even remotely possible. In this situation, it is my intention to give an introductory overview over theoretical approaches that have been used in the past and that are still being pursued. In doing so, I confine myself to a discussion of the basic underlying ideas and concepts, and I attempt in a simple fashion to sketch the theoretical framework and the mathematical tools that stand at the beginning of a detailed implementation. The further development of a given approach, the mathematical technicalities, that eventually lead to cross sections or other experimentally verifiable numbers are largely omitted. Final theoretical results are given, if they can be represented in a simple form. A comparison with experimental data is provided only for a limited number of selected examples. The reader is referred to the cited literature, which is certainly far from being complete but which should furnish a suitable starting point for further search.

The present lectures concentrate on atomic collisional processes, for simplicity mainly illustrated for prototype single-electron systems, and they leave the consideration of atomic structure and many-electron effects to other reviews. It is assumed that the reader is familiar with nonrelativistic quantum mechanics and that he had an introduction to the Dirac theory for relativistically moving electrons.

The basic processes to be considered in Part I of this overview are excitation, ionization and charge transfer. The overview is organized in the following manner: After introducing some basic concepts in Chapter 1, I consider slow ion-atom collisions in Chapter 2, which are best treated in a basis expansion, that is, we expand the exact wavefunction in terms of a suitably chosen set of atomic or molecular basis states. The accuracy of the solution then depends essentially on the choice of the basis set and on the finite number of states that can be handled numerically. In order to bear out the main features, I discuss some simple two-state systems and then proceed to a realistic treatment. In a mathematical appendix to this Chapter, the main standard basis sets are discussed. In Chapter 3, I am dealing with fast collisions, for which the basis expansion fails and which can be suitably treated by perturbative methods. For these methods, a general outline is given as well as applications to direct reactions, i.e., to excitation and ionization. In Chapter 4, the qualitatively different treatment of charge transfer is discussed, a rearrangement collison, in which initial and final states obey different Hamiltonians and hence require particular care. It turns out to be crucial to incorporate the long range of the Coulomb interaction from the outset. Up to here, nonrelativistic collisions are considered.

In Part II, from Chapter 5 on, attention is turned to relativistic collisions. If properly generalized, most of the methods discussed before can be adopted. Some of the problems pertaining also to nonrelativistic collisions are discussed here in more detail. As an introduction to relativistic collisions, I treat relativistic kinematics in Chapter 5 and discuss the fields of relativistically moving charges, whose space-time structure constitutes the main difference to nonrelativistic collisions. In Chapter 6, the reader is reminded of the quantum mechanics for relativistic electron motion, which occurs in high-Z ions or atoms. Chapter 7 is devoted to general features of relativistic projectile motion, in particular to effects caused by the long range of the Coulomb potential. Perturbative and non-perturbative methods are discussed with the aim of solving the time-dependent two-center Dirac equation, which describes relativistic ion-atom collisions involving high charges Z. The specialization to excitation and ionization is presented in Chapter 8. Subsequently, in Chapter 9, charge exchange is considered, again with particular emphasis on the influence of long-range Coulomb interactions. Chapter 10 deals with radiative electron capture (REC), a charge exchange process, in which a photon is emitted simultaneously with the transfer, thus relaxing the severe constraints on the kinematics by energy-momentum conservation. The process can essentially be understood as the inverse of the photoelectric effect. In this Chapter, the nonrelativistic and the relativistic treatments are combined. Finally, in Chapter 11, I discuss a process, namely the production of electron-positron pairs, which occurs only at sufficiently high relativistic collision energies. While the positron is always unbound, the electron may be created in a free state or, alternatively, in a bound state of the projectile. At extemely high collision energy and for close collisions, it is theoretically also possible to produce multiple electron-positron pairs.

In Part III, as a supplement, attention is focused on some selected topics that have found recent interest: In Chapter 12, I briefly introduce hyperspherical coordinates, which form a useful tool for treating three-body problems and electron correlations, in Chapter 13, a subject of much recent interest that borders to solid-state physics, namely hollow atoms in micro-capillaries, is discussed in a highly simplified manner. In Chapter 14, for crystalline solids, the spatial periodicity of the potential produced by the lattice atoms is felt as a temporal periodicity by penetrating ions and can lead to resonant coherent excitation of atomic states. Finally atomic physics with antiprotons is briefly outlined in Chapter 15, in particular the unexpectedly long-lived antiprotonic states after capture into helium. All these subjects would deserve detailed studies, but, for a simple introduction, I confine myself to the first steps.

I am grateful for the long-standing collaboration with many colleagues, in particular W.E. Meyerhof, D.P. Dewangan, N. Toshima, A. Ichihara and Th. Stöhlker. In the course of years, I also benefited from fruitful cooperation with H. Narumi, T. Ishihara, T. Shirai, A. Salop, A. Belkacem, D.C. Ionescu, V.M. Shabaev, V.A. Yerokhin, A.N. Artemyev, T. Brunne, and Y. Yamazaki. While working on the Lectures, I received indispensable help by W. Fritsch with all kinds of questions and problems. The lectures developed from a course given during a 2004 Summer School at the Institute of Modern Physics, Chinese Academy of Sciences, Lanzhou, China. I greatly appreciate the hospitality extended to me by X. Ma and his colleagues.

Berlin   Jörg Eichler
June 2005

x

Acknowledgement

Certain figures in this book are based on diagrams presented in published research articles, identified in the captions. Permission for the use of this material, granted by the publishers and by (at least one) of the authors of the articles concerned, is gratefully acknowledged. In particular, I am indebted to authors permitting me to reproduce their unpublished material, in particular to T. Brunne, K. Hencken, M. Hoshino, A. Ichihara, D.C. Ionescu, R. Morgenstern, and Th. Stöhlker.

Contents

Part I

Nonrelativistic collisions

Chapter 1

Introduction

Within the vast field of atomic physics, collisions of heavy ions with atoms define one of the most active areas of research. In the last decades, the design and construction of accelerators needed for these experiments, as well as the theoretical description of ion-atom collisions has advanced considerably. In these lectures, I want to provide an introduction into the theory aiming more at the fundamental concepts and ideas than on the mathematical and numerical implementation. For monographs on nonrelativistic and relativistic collisions, see [1, 2] and [3], respectively. These books contain more details, also of a technical nature.

In this introductory Chapter, I start with some general remarks on the qualitative behavior of atomic processes and on the units to be used, and then proceed stepwise from the most general quantum formulation for a many-electron system to a semiclassical single-electron system, which is amenable to a theoretical treatment and forms the starting point for the discussions presented here. On the way, we show that the nuclear motion can be described by classical mechanics defining trajectories, that are, to a good approximation, independent of the electronic processes, and, subsequently, establish a "semiclassical description", in which the nuclear motion is treated classically while the electronic motion obeys quantum mechanics. In a next step, it is then argued that for many applications, in particular to gain insight into the basic mechanisms, it is sufficient to consider a single active electron. At the end, I mention the possibility to describe also the electron motion classically, an approach that has been successful in many cases but is hard to justify.

In an ion-atom collision, an ion or atom with velocity v or energy E impinges on a target, which is usually at rest in the laboratory system. Figure 1.1 shows a classical picture of the collision between the two nuclei, one or

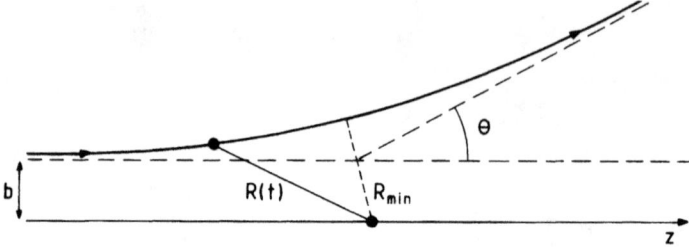

Figure 1.1: Classical trajectory of the projectile in the laboratory system

both of them carrying bound-state electrons. An important parameter for describing a classical collision is the impact parameter $\mathbf{b}$, which describes the lateral displacement of the asymptotic trajectory at time $t \to -\infty$ from the symmetry axis taken as the z-axis. The plane spanned by the z-axis and the vector $\mathbf{b}$ is denoted as the scattering plane. For structureless particles, e.g., point charges, the geometry is axially symmetric around the z-axis, so that, for example, differential cross sections are independent of the azimuthal angle.

In atomic collisions, the impact parameter b is assumed to be large enough that the nuclei do not touch or penetrate each other. This implies that the detection of the scattered projectile ensures that no nuclear reaction has taken place. The scattering angle θ refers to the asymptote at time $t \to \infty$. Figure 1.1 mainly serves to illustrate the general situation. In most applications considered here, the impact energy is so high that the deflection can be neglected, and we are dealing with a straight-line trajectory

$$\mathbf{R}(t) = \mathbf{b} + \mathbf{v}t \quad \text{with} \quad \mathbf{b} \cdot \mathbf{v} = 0. \tag{1.1}$$

In these cases, the distance of closest approach is $R_{\min} = b$.

For close encounters between point charges, the dependence of the deflection on the impact parameter is illustrated in Fig. 1.2. Note that at large impact parameters, relevant to peripheral atomic collisions, the deflection becomes negligible.

In general, we have to consider a complicated many-body system comprised of two heavy particles, the projectile nucleus P with charge number Z_P and the target nucleus T with charge number Z_T, and furthermore of $N = N_P + N_T$ electrons.

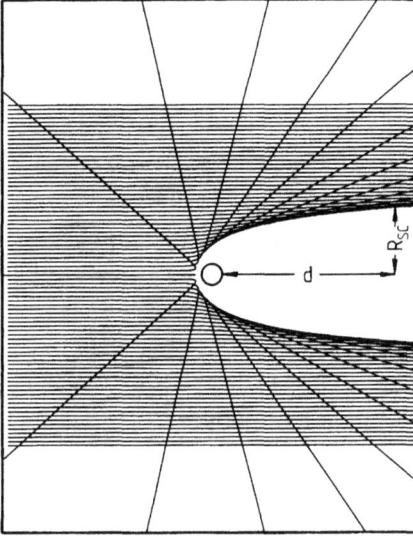

Figure 1.2: Family of trajectories in ion-ion collisions to illustrate the caustic (shadow cone). From [4].

1.1 Classification of collisions

In order to exclude nuclear reactions, we confine ourselves to peripheral collisions : The distance of closest approach $R_{\min}$ (or the impact parameter b for straight-line trajectories) has to be larger than the sum of the nuclear radii, i.e., $R_{\min} > R_P + R_T$, where the subscripts P and T refer to target and projectile, respectively.

The collision velocity v constitutes an important parameter. For *non-relativistic collisions* we have $v \ll c$, where c is the speed of light, while for *relativistic collisions* $v \approx \mathcal{O}(c)$ or at least comparable to c. Moreover, we have to distinguish between *slow* and *fast* collisions. For slow collisions, the projectile is slow compared to the electron considered. This means that the electrons are able to adjust their motion at each instant of time to the position of the projectile. That is, for slow collisions, we have

$$v \ll v_e \approx \frac{\alpha Z_T}{n} c, \qquad (1.2)$$

where $\alpha = 1/137.036$ is the fine-structure constant and n the principal quantum number of the target atom. Correspondingly, for fast collisions,

$v \gg v_e$. For relativistic collisions, this criterion has to be modified, see
Eq. 7.6 and Fig. 7.1 in Chap. 7. For the simplest or prototype atomic
process $H^+ + H$, we have the following qualitative energy dependencies:

- Low energies ($v \ll v_e$): Mainly electron transfer into the resonant
  projectile state. There is almost no energy transferred from the nuclear motion.

- High energies ($v \gg v_e$): All cross sections decrease with energy. Electron transfer is very weak compared to excitation or ionization. The
  projectile acts as a small perturbation.

- Intermediate energies (no sharp boundaries): A number of channels,
  defined by specific bound states of target or projectile or continuum
  states, are populated with comparable strength.

1.2   Units

In these lectures, we will be flexible in the usage of units : We keep *full units*,
that is, retain all physical constants explicitly in some basic equations and
in order to to help the understanding. However, often expressions become
too bulky and lengthy to be written down, in particular in exponents.
Therefore, we mostly adopt *atomic units* (a.u.) with $\hbar = 1$, $m_e = 1$, and
$e = 1$ in the nonrelativistic theory. For more accurate numbers and more
units, see Appendix A.

- Unit of length (Bohr radius):

$$a_0 = \frac{\hbar^2}{m_e e^2} = 5.29 \cdot 10^4 \text{ fm} \quad (1 \text{ fm} = 10^{-15} \text{ m})$$

- Unit of energy (Hartree):

$$\frac{e^2}{a_0} = 27.2 \text{ eV}, \quad \text{the Rydberg unit is } Ry = \frac{e^2}{2a_0} = 13.6 \text{ eV}$$

- Atomic mass unit:

$$u = 1823 \, m_e = 931.5 \text{ MeV}/c^2$$

In the relativistic theory, we generally refer to *"natural" (relativistic)
units* with $\hbar = 1$, $m_e = 1$, and $c = 1$, however, sometimes keeping m_e
explicitly.

- Unit of length (Compton wave length of the electron):

$$\lambda_c = \frac{\hbar}{m_e c} = 386 \ \text{fm}$$

- Unit of energy (rest energy of the electron):

$$m_e c^2 = 511 \ \text{keV}$$

For many applications in ion-atom collisions, the relevant length parameter is the K-shell radius. We give this quantity here in three frequently used units.

$$a_K(Z) \ = \ 1/Z \ \text{a.u.} \ = \ 5.29 \cdot 10^4/Z \ \text{fm} \ = \ 137/Z \ \lambda_c \quad (1.3)$$

For a general n-shell, we have $a_n(Z) = n \, a_K(Z)$.

1.3 Quantum formulation

As a preliminary starting point, we write down the stationary Schrödinger equation for target and projectile nuclei plus N electrons. This is, in principle, the problem we have to solve.

$$H \, \Psi(\mathbf{R}_P, \mathbf{R}_T; \mathbf{r}_1, \cdots, \mathbf{r}_N) = E \, \Psi(\mathbf{R}_P, \mathbf{R}_T; \mathbf{r}_1, \cdots, \mathbf{r}_N) \quad (1.4)$$

with

$$H \ = \ -\frac{\hbar^2}{2M_T}\nabla_T^2 - \frac{\hbar^2}{2M_P}\nabla_P^2 - \sum_{i=1}^{N}\frac{\hbar^2}{2m_e}\nabla_i^2$$

$$- \sum_{i=1}^{N}\left(\frac{e^2 Z_P}{|\mathbf{r}_i - \mathbf{R}_P|} + \frac{e^2 Z_T}{|\mathbf{r}_i - \mathbf{R}_T|}\right) + \sum_{i<j}^{N}\frac{e^2}{|\mathbf{r}_i - \mathbf{r}_j|}, \quad (1.5)$$

where M_T and M_P are the masses of the target and the projectile nuclei, respectively, and the coordinates of the projectile nucleus $\mathbf{R}_P$, of the target nucleus $\mathbf{R}_T$, and of the individual electrons $\mathbf{r}_i$ refer to a fixed origin. The scattering wavefunction describing a collision spreads over all space and is subject to the boundary condition that at $|\mathbf{R}_P - \mathbf{R}_T| \to \infty$, the electrons are shared between the collision partners in a specific way. The nuclear motion is described by plane waves (aside from a residual Coulomb distortion if both fragments are charged, see Sec. 4.3). The relative nuclear motion will have to be described by a partial-wave expansion. In a quantum formulation including the nuclear motion, energy and momentum are conserved.

In general, this problem cannot be solved, hence simplifications are necessary.

1.4   Classical description of the nuclear motion

Fortunately, it is justified to introduce drastic simplifications by treating
the nuclear motion classically. It is assumed that the nuclei follow predetermined classical trajectories, independently of the electronic reactions
taking place. Why is this possible?

- The nuclear masses are much greater than the electron mass, and their
  momenta P_Γ and P_P are much greater than the average electronic
  momenta $\langle p_e \rangle$:

$$M_P, \; M_\Gamma \gg m_e \quad \text{and} \quad P_P, \; P_\Gamma \gg \langle p_e \rangle,$$

  and the nuclear de Broglie wavelengths $\lambda = \hbar/P$ are much smaller
  than typical atomic sizes

$$\lambda_P, \; \lambda_\Gamma \ll a_0, \quad \text{in general} \quad \lambda_P, \; \lambda_\Gamma \ll \frac{a_0}{Z_{P,T}}.$$

  Hence the nuclear wave packets are very small compared to atomic
  dimensions.

- The energy ΔE transferred to the electronic system during the collision is very small compared to the kinetic energy T_P of the projectile.

$$\Delta E \ll T_P.$$

  The nuclear motion suffers essentially no energy loss and can be well
  described by an elastic predetermined trajectory, e.g., by a Ruther-
  ford trajectory parametrized by the relative internuclear velocity $\mathbf{v}_{\text{rel}}$
  and by the impact parameter b, see Figs. 1.1 and 1.2.

However, it should be noted that finer details of the reaction can be investigated by precision measurements of the nuclear recoil . This method
has been developed into momentum imaging or a "reaction microscope"
in atomic collisions, a very important tool for studying electronic momentum changes during the reaction, see, e.g. [5, 6]. In this way, kinematically
complete measurements of reactions have become possible. In the following,
we will always assume predetermined classical trajectories for the nuclear
motion.

For completeness, let me give a brief outline of particle scattering within
classical mechanics. If r is the radial projectile coordinate with respect to
the target nucleus, and φ the angle in the scattering plane with respect to

the bisector, indicated by a short-dashed line in Fig. 1.1, the trajectory is determined by two constants of the motion, the energy E and the angular momentum L perpendicular to the scattering plane.

$$E = \frac{\mu}{2}\left(\frac{dr}{dt}\right)^2 + \frac{L^2}{2\mu r^2} + V(r),$$ (1.6)

where $\mu = M_P M_T/(M_P + M_T)$ is the reduced mass of the nuclei and $V(r)$ the interaction potential. Furthermore,

$$L = \mu r^2 \frac{d\varphi}{dt}.$$ (1.7)

Solving for the derivatives dr/dt, $d\varphi/dt$, and dividing one equation by the other, we obtain

$$\frac{d\varphi}{dr} = \frac{L}{r^2 \sqrt{2\mu[E - V(r) - L^2/(2\mu r^2)]}},$$ (1.8)

which can be integrated to yield

$$\varphi(r) = \int_{R_{min}}^{r} dr' \frac{L}{r'^2 \sqrt{2\mu[E - V(r') - L^2/(2\mu r'^2)]}}.$$ (1.9)

Since

$$\Theta = \pi - 2\varphi(\infty)$$ (1.10)

and

$$L = bp = b\sqrt{2\mu E},$$ (1.11)

we obtain the classical deflection function describing the asymptotic deflection $\Theta(b)$, see Fig. 1.1, as a function of the impact parameter b in the form

$$\Theta(b) = \pi - 2\int_{R_{min}}^{\infty} dr \frac{b}{r^2 \sqrt{1 - V(r)/E - b^2/r^2}}.$$ (1.12)

Specifically, for a repulsive Coulomb potential

$$V(r) = \frac{e^2 Z_P Z_T}{r},$$

one obtains the classical Coulomb deflection function

$$\Theta(b) = 2\arctan\left(\frac{e^2 Z_P Z_T}{2bE}\right).$$ (1.13)

For high energies,

$$2E \gg \frac{e^2 Z_{\mathrm{P}} Z_{\mathrm{T}}}{b},$$

the deflection may be neglected and the curvilinear trajectory may be re-placed by a straight-line trajectory .

To conclude this Section, let us write down the differential cross section. Equation 1.12 implies that particles having impact parameters between b and $b + db$ are scattered into the angular interval between θ and $\theta + d\theta$ (Note that $\Theta(b)$ is the deflection function while θ is the angular coordinate). The differential cross section $d\sigma/d\Omega$ is defined as the number of particles deflected by the angle θ into an element of solid angle $d\Omega = 2\pi \sin\theta \, d\theta$, per unit time and unit incident flux. The fraction of the total flux passing through the annular area between b and $b + db$ is given by $2\pi b \, db$, where $db = (db/d\Theta) \, d\theta$ corresponds to the angular interval $d\theta$, so that

$$\frac{d\sigma}{d\Omega} = \frac{2\pi b \, d\theta}{2\pi \sin\theta \, d\theta} \left| \frac{db}{d\Theta} \right| = \frac{b}{\sin\theta \, |d\Theta/db|} \tag{1.14}$$

In this way, we readily obtain the Rutherford cross section from Eq. (1.13) as

$$\frac{d\sigma}{d\Omega} = \left(\frac{e^2 Z_{\mathrm{P}} Z_{\mathrm{T}}}{4E} \right)^2 \frac{1}{\sin^4(\theta/2)}. \tag{1.15}$$

Coulomb trajectories are often used for low-energy collisions, however, for high energies we always assume straight-line trajectories.

1.5   Semiclassical description

Having justified that in an energetic collision the projectile follows a prede-termined classical trajectory $\mathbf{R}(t)$, independently of all electronic processes, we have to consider a time-dependent Schrödinger equation for the electron motion. Because for a given energy this description is characterized by the impact parameter b, this approach is also denoted as *impact-parameter pic-ture* [†]. Referring all coordinates to the target nucleus, the time-dependent Schrödinger equation replacing Eq. (1.4) takes the form

$$\left(H(t) - i\frac{\partial}{\partial t} \right) \Psi(\mathbf{r}_1, \cdots, \mathbf{r}_N; t) = 0, \tag{1.16}$$

[†]For a derivation from the stationary quantum formulation see, e.g. [3]

where

$$H \;=\; -\sum_{i=1}^{N} \frac{\hbar^2}{2m_e} \nabla_i^2 - \sum_{i=1}^{N} \left(\frac{e^2 Z_P}{|\mathbf{r}_i - \mathbf{R}(t)|} + \frac{e^2 Z_T}{|\mathbf{r}_i|} \right) + \sum_{i<j}^{N} \frac{e^2}{|\mathbf{r}_i - \mathbf{r}_j|}.$$

$$(1.17)$$

Here, the first term represents the kinetic energy of the electrons, the second term the interaction of the electrons with the projectile and target nucleus, respectively, and the third term the interelectronic interaction. In contrast to Eq. (1.5), we here have to consider only electronic coordinates as dynamical variables. From the assumption of predetermined nuclear trajectories, it follows that energy and momentum are *not* formally conserved in the impact parameter picture.

1.6 Single-electron approximation

The set of equations (1.16) and (1.17) is still quite complicated. For many processes and in order to gain insight into the basic mechanisms, it is justified to consider a *single active electron*. This is particularly true, if one considers inner-shell electrons, for which the electron-nucleus interaction in heavy atoms is much larger than the electron-electron interaction. We hence assume that either there are no other electrons, or we can consider them as inert, i.e., not taking part in the reaction. The problem then is reduced to a time-dependent three-body system consisting of two nuclei and one electron.

Within the nonrelativistic domain, we have to solve a time-dependent single-electron problem following from Eq. (1.16)

$$\left(H(t) - \mathrm{i}\frac{\partial}{\partial t} \right) \Psi(\mathbf{r}, t) = 0,$$

$$(1.18)$$

with

$$H(t) = -\frac{\hbar^2}{2m_e} \nabla^2 - \frac{e^2 Z_T}{r_T} - \frac{e^2 Z_P}{r_P(t)}.$$

$$(1.19)$$

Assuming that the target is located at the origin, $\mathbf{r}_T = \mathbf{r}$, and $\mathbf{r}_P(t) = \mathbf{r} - \mathbf{R}(t)$ is the electron's explicitly time-dependent position with respect to the projectile.

1.7   Classical-trajectory Monte Carlo calculations (CTMC)

If one describes not only the nuclear motion but also the electronic motion by classical trajectories, one is led to the "classical-trajectory Monte-Carlo" method, which has been first introduced by Abrines and Percival [7], subsequently applied by Olson and Salop [8] and later by Olson and collaborators in a great number of publications. In fact, a vast literature exists, which treats the electron motion classically, irrespective of the quantum nature and the value of the de Broglie wave length.

    These calculations start from Hamilton's equations

$$\frac{\mathrm{d}\mathbf{r}_i}{\mathrm{d}t} = \frac{\partial H}{\partial \mathbf{p}_i}$$

$$\frac{\mathrm{d}\mathbf{p}_i}{\mathrm{d}t} = -\frac{\partial H}{\partial \mathbf{r}_i}, \tag{1.20}$$

where the label $i = 1, 2$ denotes the Jacobi coordinates $\mathbf{r}_T$ and $\mathbf{R}$, while $\mathbf{p}_i$ are the canonical conjugate momenta, see, e.g. [9].

    For a one-electron problem, the initial condition at $t \to -\infty$ is given, say, by a hydrogen-like atom. This atom is simulated classically by adopting a microcanonical (normalized) momentum distribution (see [1], Appendix 3.1), which is known to be identical to the quantum distribution. This distribution is valid for the 1s shell or, in general, for a *complete* principal shell n.

$$\rho(\mathbf{p}) = \frac{8p_0^5}{\pi^2} \frac{1}{(p^2 + p_0^2)^4} \quad \text{with} \quad p_0 = \frac{Z}{n} \text{ (a.u.)} \tag{1.21}$$

The classical density distribution for the hydrogenic 1s shell in position space is

$$\rho(\mathbf{r}) = \begin{cases} \frac{Z^3}{2\pi^2} r^{-1/2} \left(\frac{2}{Z} - r\right)^{1/2} & \text{if } r \leq \frac{2}{Z} \\ 0 & \text{if } r > \frac{2}{Z} \end{cases} \tag{1.22}$$

or, integrated over a spherical shell,

$$\rho(r) = \int \rho(\mathbf{r})r^2 \mathrm{d}\Omega = \begin{cases} \frac{2Z^3}{\pi} r^{3/2} \left(\frac{2}{Z} - r\right)^{1/2} & \text{if } r \leq \frac{2}{Z} \\ 0 & \text{if } r > \frac{2}{Z} \end{cases} \tag{1.23}$$

This distribution is not normalized – in contrast to (1.21) – and has a sharp cut-off, which is clearly not in agreement with the exponential tail arising from quantum theory. Therefore, in cases of distant collisions, when the exponential tail of the wavefunction matters, one may expect unrealistic results.

For the *final state* at $t \to \infty$, one has to distinguish the various possibilities: (a) if the electron remains near the target, one has an elastic collision or *excitation*, (b) if the electron follows the projectile nucleus (and has about the same momentum), we have *electron transfer*, (c) if the electron is far away from both target and projectile, one interprets this as *ionization*. Of course, in order to calculate cross sections, one has to generate a large number, millions of initial conditions, randomly, by a Monte-Carlo method and then follow the classical trajectories subject to the equations (1.20), until the nuclei have reached a large separation in the outgoing channel. In this way, while a theoretical foundation is still lacking, the method has produced good results in many cases, provided that one confines oneself to one-electron systems.

One may conjecture that one reason for the success of the CTMC method lies in the dynamic symmetry of the Coulomb or Kepler problem, which in a classical description leads to an additional constant of the motion, the Lenz-Runge vector. In a quantum formulation this corresponds to the well-known $O(4)$ symmetry of the hydrogen problem [10] responsible for the "accidental degeneracy" within a principal shell. As a further consequence of this dynamical symmetry, the Rutherford cross section is obtained by a classical as well as by a quantum description. Hence, apparently in this basic case of an unbound Coulomb system, a classical description leads to the same result as a quantum formulation.

As is well known, it is not possible, within classical physics, to stabilize a two-electron system like the helium atom. This, together with the instability of the hydrogen atom with respect to the emission of radiation, was at the origin of Bohr's quantum theory. Nevertheless, CTMC is often applied to multielectron systems by simply avoiding to wait for the system to decay. Correspondingly, the results are not always reliable. In the present review, we will refrain from further discussing the CTMC method.

Chapter 2

Low-energy collisions: Basis expansions

In this Chapter, we assume that the relative velocity v of the collison partners is small compared to the speed v_e of the electrons involved, or at least comparable to it. For the kinetic energy T_P per atomic mass unit u, this corresponds to

$$0.2\sqrt{T_P(\text{keV/u})} \ll Z/n, \tag{2.1}$$

where n is the principal shell of the bound electron. If this is satisfied, the electrons are fast enough to adjust their motion to the relative position of the collision partners at each instant of time. In this way, a "quasimolecule" is formed, in which the the internuclear separation $R(t)$ is a slowly varying function of time. In a collision with a fixed target, the trajectory of the projectile is described by $\mathbf{R}(v, \mathbf{b}, t)$, which – depending on the velocity – follows a curved orbit, see Fig. 1.1, or a straight-line trajectory $\mathbf{R} = \mathbf{b} + \mathbf{v}t$ (1.1) for sufficiently high kinetic energy. In general, the Hamiltonian, in atomic units, is given by

$$H(\mathbf{r}, \mathbf{R}) = -\tfrac{1}{2}\nabla^2 + V_P\big(\mathbf{r}_P(t)\big) + V_T(\mathbf{r}_T), \tag{2.2}$$

where we denote $\mathbf{r}_T = \mathbf{r}$ and $\mathbf{r}_P(t) = \mathbf{r} - \mathbf{R}(t)$.

While there are many ways to describe a slow collision approximately [1], we concentrate in Sec. 2.1 on introducing the coupled-channel method, which is, in principle, exact with its accuracy being limited only by the size of the basis set used for the expansion of the exact wavefunction. If electron transfer is to be included in the description, the expansion should comprise target- as well as projectile-centered states. After a general outline of the approach, we turn in Sec. 2.2 to the modification needed to account for the motion of the projectile, that is, the electron translation factors.

For very slow collisions, the motion of the nuclei and of the electrons essentially separate as described in Sec. 2.3 by the Born-Oppenheimer approximation. Under this condition, see Sec. 2.4, projectile and target form a quasimolecule, whose energy levels are given by a correlation diagram as a function of the time-dependent internuclear separation. Transitions occur at curve crossings, when the energy levels come close to each other. During the passage of the system through the reaction zone, quasimolecular x-rays may be emitted (Sec. 2.5). In Sec. 2.6, we discuss the idealized situation involving essentially only two levels, giving rise to simplifications and to the Landau-Zener approximation, which sometimes gives useful estimates. Subsequently, in Sec. 2.7, we illustrate an example of Stückelberg oscillations, originating from interferences between reaction amplitudes arising from the incoming and outgoing passage of the projectile.

In quantitative dynamical calculations, Sec. 2.8, the passage of the projectile through the reaction zone represented in the correlation diagrams leads to radial couplings and further to rotational couplings induced by the rotation of the internuclear line during the collision. In Sec. 2.9, for faster collisions, it is most practical to adopt a two-center atomic expansion within a coupled-channel description. Some results of this approach are compared with experimental data. Finally, in a mathematical appendix, Sec. 2.10, we present various basis sets that have been used for expanding the exact wavefunctions.

2.1   Coupled-channel description: General outline

In the coupled-channel description or in close-coupling calculations, we expand the exact wavefunction $\Psi(\mathbf{r}, t)$ in terms of a suitably chosen basis set composed of N basis functions $\psi_k(\mathbf{r}, t)$ (see e.g. [1, 2] and the excellent review article by Fritsch and Lin [11]) in the form

$$\Psi(\mathbf{r}, t) = \sum_{k=1}^{N} a_k(t)\psi_k(\mathbf{r}, t). \tag{2.3}$$

The quantities $a_k(t)$ are (unknown) complex amplitudes for occupying the state $\psi_k(\mathbf{r}, t)$. These amplitudes have to be determined by inserting the expansion (2.3) into Eq. (2.2) and thus finding an optimal solution within the given basis set.

Various different basis sets $\{\psi_k\}$ have been used, see Sec. 2.10. In general, it is desirable, that the basis sets satisfy certain requirements:

1. The treatment becomes particularly convenient in an atomic-orbital (AO) expansion, that is, if the basis states are eigenstates of H_T and, in the case of transfer, basis states of H_P are also included, see Eq. (2.13). This allows for the specification of initial conditions and for the direct physical interpretation of the final amplitude. The truncation to a maximum value of N is usually dictated by the available computer power.

2. For an AO expansion, the basis should include the initial and final states as well as states that are strongly coupled.

3. In principle, a primary basis is completely arbitrary. This freedom has to be paid for by a sufficiently large size of the set if it is not composed of physical states. However, for the purpose of physical interpretation and for the formulation of initial conditions, one usually constructs a secondary ("contracted") basis of approximate eigenstates by diagonalizing the relevant atomic Hamiltonians. Resulting unphysical states in the continuum may be discarded or sometimes kept as "pseudostates" in order to approximately represent the continuum. From there on, one proceeds according to (1) and (2).

4. If a nonphysical primary basis is used, it should at least lend itself for a convenient evaluation of the matrix elements. Sometimes the convenience of evaluation, e.g., by using a Gaussian basis (GTO) or Slater-type orbitals (STO) or a Sturmian basis set, see Sec. 2.10, is the dominant criterion for choosing a basis set.

We now assume that the $\{\psi_k\}$ are physical eigenstates of target or projectile obtained by diagonalizing the stationary atomic Hamiltonians. By construction, they are orthogonal and normalized at $t \to \pm\infty$. Let $\psi_i(\mathbf{r}, -\infty)$ and $\psi_f(\mathbf{r}, \infty)$ represent initial and final states. Then, for an electron initially occupying the state i, the *initial conditions* for the probability amplitudes $a_k(t)$ are

$$
\begin{aligned}
a_i(-\infty) &= 1 \quad (\text{or } e^{i\alpha}, \ \alpha \text{ real}) \\
a_k(-\infty) &= 0, \quad k \neq i.
\end{aligned}
$$

(2.4)

The transition probability from the initial to the final state for a given trajectory $\mathbf{R}(v, \mathbf{b}, t)$ of the projectile is

$$
P_{i\to f}(v, \mathbf{b}) = |a_f(+\infty)|^2.
$$

(2.5)

The total cross section is then obtained by integrating over all impact parameters

$$
\sigma_{i\to f}(v) = 2\pi \int_0^\infty P_{i\to f}(v, \mathbf{b}) \, b \, db.
$$

(2.6)

The problem of calculating the cross section is now reduced to the determination of a finite set of amplitudes $a_k(t)$. This is achieved by requiring that Ψ must obey the time-dependent Schrödinger equation within the space of basis functions, that is

$$\left\langle \psi_j \left| \mathrm{i}\frac{\partial}{\partial t} - H(\mathbf{r},\mathbf{R}) \right| \Psi \right\rangle = 0, \qquad (2.7)$$

where the Hamiltonian is given by Eq. (2.2). Inserting the expansion (2.3) into (2.7) and integrating over the electronic coordinates, we obtain a set of coupled linear differential equations for the amplitudes $a_k(t)$:

$$\sum_{k=1}^{N} N_{jk}(t)\,\dot{a}_k(t) = -\mathrm{i}\sum_{k=1}^{N} M_{jk}(t)\,a_k(t), \qquad j = 1,\cdots,N, \qquad (2.8)$$

where the dot denotes the time derivative,

$$N_{jk}(t) = \langle \psi_j | \psi_k \rangle \quad \text{and} \quad M_{jk}(t) = \left\langle \psi_j \left| \mathrm{i}\frac{\partial}{\partial t} - H \right| \psi_k \right\rangle \qquad (2.9)$$

are the overlap and coupling matrix elements, respectively. The inclusion of the overlap matrix in Eq. (2.8) permits the use of general nonorthogonal basis states. The equations (2.8) can be written conveniently in a more compact way as

$$\mathrm{i}\,\dot{\mathbf{a}} = \mathbb{N}^{-1}\,\mathbb{M}\mathbf{a}, \qquad (2.10)$$

where $\mathbf{a}$ is the column vector composed of the amplitudes a_k, while $\mathbb{N}$ and $\mathbb{M}$ are the overlap and coupling matrices, respectively.

In principle, for an infinite basis set, the $a_k(t)$ would provide an exact solution of the Schrödinger equation. The approximation consists in truncating the basis set to a finite value of N. All physical assumptions are contained in the proper choice of the basis set $\{\psi_k\}$ and in the initial conditions. A "good" basis, that is a basis that takes into account as much physics of the system as possible, is expected to yield good convergence (with respect to the number N of states to be included). In other words, if one starts with a "good" basis, one needs a smaller number of basis states to obtain numerical convergence.

The solution of the coupled equations preserves unitarity (see, e.g., [11]), even for a truncated basis, as long as further approximations in the evaluation of the matrix elements are avoided. This means that

$$\frac{\mathrm{d}}{\mathrm{d}t}\langle \Psi | \Psi \rangle = 0 \qquad \text{or} \qquad \sum_k |a_k(\infty)|^2 = 1, \qquad (2.11)$$

taking into account Eq. (2.4). Equation (2.11) provides a useful test for checking the accuracy of numerical calculations.

For many practical applications, it is convenient or even necessary to specify the general expansion (2.3) as an expansion in terms of time-dependent target and time-dependent projectile wavefunctions

$$\Psi(\mathbf{r}, t) = \sum_k a_k^{\mathrm{T}}(t)\, \psi_k^{\mathrm{T}}(\mathbf{r}_{\mathrm{T}}, t) + \sum_k a_k^{\mathrm{P}}(t)\, \psi_k^{\mathrm{P}}(\mathbf{r}_{\mathrm{P}}, t), \qquad (2.12)$$

see Sec. 2.9 for details. Here, it is important to realize that the projectile, in general, is moving along an arbitrary trajectory. However, for simplicity, let us assume that the target nucleus is located at the origin and the projectile follows a straight-line trajectory $\mathbf{R}(t) = \mathbf{b} + \mathbf{v}t$ with $\mathbf{v} = \mathrm{const}$. This is a good approximation for sufficiently high collision energies. In fact, even if asymptotically the projectile is deflected by a finite small angle, it is justified for the calculation of matrix elements and cross sections to adopt a rectilinear path within the reaction zone.

In order to construct the desired wavefunctions $\psi_k^{\mathrm{T}}(\mathbf{r}_{\mathrm{T}}, t)$ and $\psi_k^{\mathrm{P}}(\mathbf{r}_{\mathrm{P}}, t)$, we first define the the space wavefunctions ϕ_k^{T}, ϕ_k^{P} of target and projectile, respectively, subject to the eigenvalue equations

$$\begin{aligned} H_{\mathrm{T}}\, \phi_k^{\mathrm{T}}(\mathbf{r}) &= \epsilon_k^{\mathrm{T}}\, \phi_k^{\mathrm{T}}(\mathbf{r}) \\ H_{\mathrm{P}}\, \phi_k^{\mathrm{P}}(\mathbf{r}_{\mathrm{P}}) &= \epsilon_k^{\mathrm{P}}\, \phi_k^{\mathrm{P}}(\mathbf{r}_{\mathrm{P}}) = \epsilon_k^{\mathrm{P}}\, \phi_k^{\mathrm{P}}(\mathbf{r} - \mathbf{R}(t)), \end{aligned} \qquad (2.13)$$

where the coordinates are denoted as in Eq. (2.2) and the one-electron separated-atom Hamiltonians are given by

$$\begin{aligned} H_{\mathrm{T}} &= -\tfrac{1}{2}\nabla^2 + V_{\mathrm{T}}(\mathbf{r}) = -\tfrac{1}{2}\nabla^2 - \frac{Z_{\mathrm{T}}}{r} \\ H_{\mathrm{P}} &= -\tfrac{1}{2}\nabla^2_{r_{\mathrm{P}}} + V_{\mathrm{P}}(\mathbf{r}_{\mathrm{P}}) = -\tfrac{1}{2}\nabla^2 - \frac{Z_{\mathrm{P}}}{r_{\mathrm{P}}(t)}. \end{aligned} \qquad (2.14)$$

Here, the equations on the right-hand side are specified for the Coulomb potentials of bare nuclei. In the following, we have to study the time-dependent solutions of the Hamiltonians (2.14).

2.2   Electron translation factors

While the target atom (at rest) obeys the time-dependent Schrödinger equation

$$\left(\mathrm{i}\frac{\partial}{\partial t} - H_{\mathrm{T}} \right) \phi_k^{\mathrm{T}}(\mathbf{r})\, \mathrm{e}^{-\mathrm{i}\epsilon_k^{\mathrm{T}} t} = 0, \qquad (2.15)$$

the projectile ion is described by an equation that takes its motion into account, embodied in the implicit time dependence of the coordinate $\mathbf{R}$. The time derivative then also acts on the coordinate $\mathbf{r}_P$, leading to

$$\left(i\frac{\partial}{\partial t} - H_P\right)\phi_k^P(\mathbf{r}_P)\,e^{-i\epsilon_k^P t} = -i\mathbf{v}\cdot\nabla_{\mathbf{r}_P}\,\phi_k^P(\mathbf{r}_P)\,e^{-i\epsilon_k^P t}. \tag{2.16}$$

Hence $\phi_k^P(\mathbf{r}_P)$ is not a stationary solution except in the coordinate system moving with the projectile. In fact, the electron travelling with the projectile carries an additional momentum and an additional energy. To take this into account, the time-dependent wavefunctions of bound states for target and projectile at infinite separations are defined, respectively, as

$$\begin{aligned}
\hat{\phi}_k^T(\mathbf{r},t) &= \phi_k^T(\mathbf{r})\,e^{-i\epsilon_k^T t}\\
\hat{\phi}_k^P(\mathbf{r}_P,t) &= \phi_k^P(\mathbf{r}_P)\,e^{i\mathbf{v}\cdot\mathbf{r}-\frac{i}{2}v^2 t}\,e^{-i\epsilon_k^P t}.
\end{aligned} \tag{2.17}$$

With the additional phase factor $e^{i\mathbf{v}\cdot\mathbf{r}-\frac{i}{2}v^2 t}$, one obtains in both cases

$$\left(i\frac{\partial}{\partial t} - H_{T,P}\right)\hat{\phi}_k^{T,P}(\mathbf{r}_{T,P},t) = 0. \tag{2.18}$$

The crucial phase factor for a projectile moving with speed $\mathbf{v}$,

$$F = \exp\!\left(i\mathbf{v}\cdot\mathbf{r} - i\tfrac{1}{2}v^2 t\right), \tag{2.19}$$

is denoted as *electron translation factor* (ETF). It takes into account momentum and energy of electron orbitals moving with the projectile. This important feature has been first introduced by Bates and McCarroll in 1958 [12]. If the translation factors are neglected, one obtains incorrect results, even at comparatively low collision velocities. The adoption of ETF in conjunction with the choice of the basis sets has become a common practice.

In all basis expansions involving states attached to a moving center, the basis functions ψ_k should be selected such that

$$\lim_{t\to\pm\infty}\psi_k(\mathbf{r},t) = \hat{\phi}_k^{T,P}(\mathbf{r},t). \tag{2.20}$$

In this case, the physically relevant asymptotic amplitudes are given by

$$\begin{aligned}
a_i(-\infty) &= \langle\hat{\phi}_i(-\infty)\,|\,\Psi(-\infty)\rangle\\
a_f(\infty) &= \langle\hat{\phi}_f(\infty)\,|\,\Psi(\infty)\rangle.
\end{aligned} \tag{2.21}$$

The phase factor $\exp(-\frac{i}{2}v^2 t)$ reflecting the kinetic energy of the moving electrons depends on time only and therefore can be absorbed in the occupation amplitude $a_k(t)$. Merely the space-dependent phase arising from the linear momentum plays a role in coupled-channel calculations.

Up to here, we have adopted linear combinations of atomic orbitals (LCAO), see Sec. 2.9. In this case, the translation factors are well defined throughout the collision. However, sometimes one uses an expansion in terms of molecular orbitals (MO), which for a fixed time, that is, for a fixed internuclear separation are eigenstates of (2.2). This implies that translation factors are uniquely defined only at asymptotic separations, see Sec. 2.8.

In either case, it is customary to factorize the basis functions in the form

$$\psi_k(\mathbf{r}, t) = \chi_k(\mathbf{r}, t)\, f(\mathbf{R}, \mathbf{r})\,, \tag{2.22}$$

where the $\chi_k(\mathbf{r})$ represent either single-center atomic orbitals or molecular orbitals with the requirement that, asymptotically,

$$\lim_{R \to \infty} f(\mathbf{R}, \mathbf{r}) = e^{i\mathbf{v}\cdot\mathbf{r}}. \tag{2.23}$$

The form of the factor $f(\mathbf{R}, \mathbf{r})$ at finite internuclear separations is left open at this point. It becomes important in dynamical calculations, for which certain prescriptions have been developed, see the discussion in Sec. 2.8.

2.3 Molecular orbital model: The Born-Oppenheimer expansion

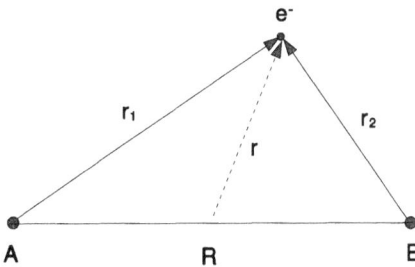

Figure 2.1: Coordinate system for a quasimolecule composed of a single electron e^- and the two nuclei A and B.

In order to give specific examples for the general theory outlined in Sec. 2.1, let us first consider very slow collisions. At low nuclear velocities, the electrons are able to adjust their motion adiabatically to the slowly rotating internuclear axis $\mathbf{R}$, see Fig. 1.1, in such a way that adiabatic molecular states $\chi_{\mathbf{R}}^{(k)}(\mathbf{r})$ with a definite angular momentum around the internuclear axis are formed. The relevant coordinate system is illustrated in Fig. 2.1. These states depend parametrically on $\mathbf{R}$, so that the functions $\chi_{\mathbf{R}}^{(k)}$ are determined for a fixed value of $\mathbf{R}$. The *Born-Oppenheimer expansion* [13] now decomposes the total wavefunction

$$\Psi(\mathbf{r}, \mathbf{R}(t)) = \sum_k \chi_{\mathbf{R}}^{(k)}(\mathbf{r}) \, F_k(\mathbf{R}(t)) \tag{2.24}$$

into terms $\chi_{\mathbf{R}}^{(k)}(\mathbf{r})$ describing the electronic motion, and terms $F_k(\mathbf{R}(t))$ describing the nuclear motion, either in a semiclassical or in a fully quantal treatment [13]. For a fixed value of $\mathbf{R}$, this expansion is exact. The *Born-Oppenheimer approximation* consists in retaining a single term in the expansion (2.24). Within this framework, the coupling between the nuclear and the electronic motion is treated as a perturbation.

In the Born-Oppenheimer approximation, electronic transitions occur between transient molecular or *quasimolecular* states. The quasimolecular excitation mechanism of inner-shell electrons, first proposed by Fano and Lichten [14] in 1965, has attracted much attention at a time when heavy-ion accelerators first became available for atomic collisions.

Imagine two atoms or ions approaching each other. When the atoms form a (quasi-) molecule in this manner, an inner-shell electron may experience an increase of its asymptotic (separated-atom-) principal quantum number during the collision, i.e., it may be raised into a higher molecular orbital (MO). The pre-collision levels of these "promoted" electrons are no longer fully occupied but contain vacancies after the collision. As a result, an "inner-shell excitation" has occurred. Since the promotion of electrons can happen only into initially vacant levels, we may say, in a many-electron picture, that during the collision vacancies from outer shells are "transferred" into inner shells.

2.4 Molecular orbitals: Correlation diagrams and couplings

The process of promotion or quasimolecular excitation is governed by "correlation diagrams", defined in the coordinate system of Fig. 2.1. A single electron moves in the field of two nuclei, A and B separated by the distance R. By solving the effective (for a many-electron system) static two-center problem (2.25), one obtains a sequence of (quasi-) molecular levels in dependence of the internuclear separation R. A basic correlation diagram

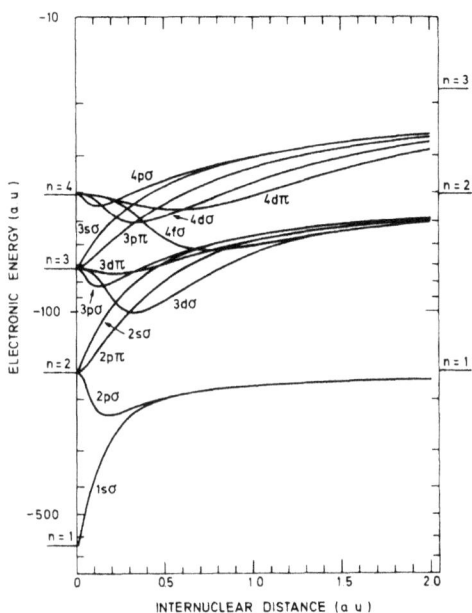

Figure 2.2: Correlation diagram for a single electron in the field of two un-screened Ar nuclei. The notation at individual curves indicates the subshell in the united-atom limit and the angular momentum projection m on the internuclear line. The notation σ, π, $\cdots$ corresponds to $m = 1, 2, \cdots$. From [15].

is provided by the energy levels as a function of R for an electron in the two-center Coulomb field of two bare nuclei. Figure 2.2 illustrates such a diagram for the homonuclear system of Ar^{17+} – Ar^{18+}. It is seen that the levels are highly degenerate both in the united- and in the separated atom limits. This degeneracy is partly removed, if the active electron is subject to screened Coulomb fields produced by additional electrons, see Fig. 2.3. Clearly, in the case of a heteronuclear system, the degeneracy in the separated-atom limit is further lifted.

This diagram allows us to illustrate the quasimolecular excitation mechanism in more detail. Let us consider an electron residing asymptotically in the Ar 1s shell and following a $2p\sigma$ orbital as the nuclei approach each other. At very small internuclear separations, this level becomes almost degenerate with the $2p\pi$ level so that the electron may be transferred by

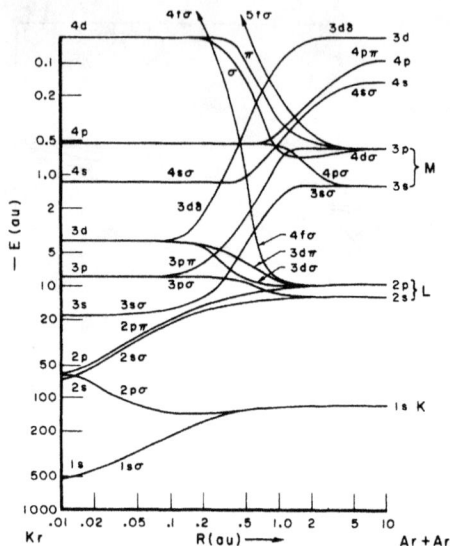

Figure 2.3: Semi-quantitative MO energy level diagram for the Ar_2 molecule as proposed by Fano and Lichten [14]. For the notation, see Fig. 2.2. Note the strong promotion of the $4f\sigma$ level. This is a diabatic diagram in the sense that avoided crossings, see Fig. 2.7, are replaced by real crossings.

rotational (Coriolis) coupling (for details, see Sec. 2.8) into this level, provided this level is empty. In this way, a K-vacancy is created after the atoms separate. In other words, a vacancy initially present in the 2p shell is "injected" into the 1s shell after separation.

Since the diagram of Fig. 2.3 is a qualitative one, it is desirable to construct more realistic correlation diagrams for experimentally available projectile-target combinations. Within a multi-electron quasimolecule, the active electron is subject to an effective single-particle Hamiltonian

$$H(\mathbf{r}, \mathbf{R}) = -\tfrac{1}{2}\nabla^2 + V^{\mathrm{eff}}(\mathbf{r}_1, \mathbf{r}_2, \mathbf{R}), \qquad (2.25)$$

where it is meaningful to drop the distiction between target and projectile. According to Fig. 2.1, the position of the electron with respect to a fixed origin (e.g. the midpoint between the nuclei) is given by $\mathbf{r}$, while $\mathbf{r}_1$ and $\mathbf{r}_2$ are the coordinates of the electron with respect to the nuclei A and B.

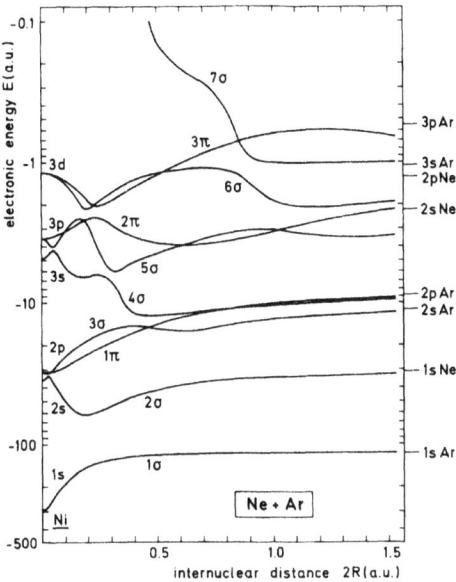

Figure 2.4: Correlation diagram for the asymmetric system Ne + Ar, calculated from an effective Thomas-Fermi potential [16]. The calculation has been restricted to $\sigma(m = 0)$ and $\pi(m = 1)$ states. From [17].

In the original publications, [14], see also [18], correlation diagrams have been drawn on a qualitative basis as in Fig. 2.3. In order to obtain realistic diagrams without solving the full many-body problem, e.g., by Hartree-Fock calculations [19], Eichler and Wille [16, 17] have constructed a simple effective potential V^{eff}, which takes into account the R-dependent screening exerted on the active electron by the interpenetrating electron clouds (Variable Screening Model). By solving the problem in prolate spheroidal coordinates $\xi = (r_1 + r_2)/(2R)$ and $\eta = (r_1 - r_2)/(2R)$, a large number of semi-quantitative correlations diagrams have been obtained and discussed, see also [20]. Examples are shown in Figs. 2.4 and 2.5. Note that energy levels with different m-values, e.g., σ and π, cross each other because they correspond to different eigenvalue problems. A change in notation should be pointed out: While, for historical reasons, in Figs. 2.2 and 2.3 the united-atom angular momentum l is used for specifying a molecular state, although for $R > 0$ the angular momentum l is no longer a good

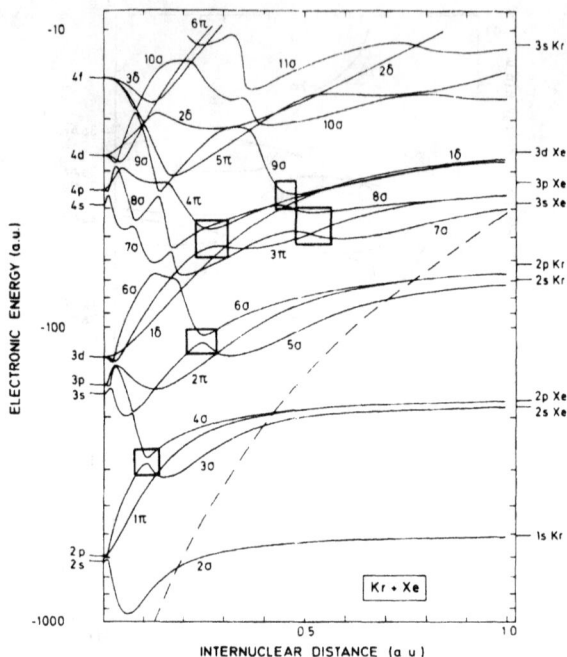

Figure 2.5: Correlation diagram for the Kr + Xe system. The closest and most localized pseudocrossings, *not* predicted by the Barat-Lichten rule [21] are indicated by boxes. The dashed line gives the height of the potential barrier between the two nuclei. From [15].

quantum number, the proper notation, used in Figs. 2.4 and 2.5, simply counts the levels with a given m-value from bottom to top.

Let us summarize the quasimolecular excitation mechanism: During a (slow) collision, the electrons follow a potential line and may experience one of the following fates:

- *Adiabatic collisions*: The electron (or a vacancy) entering the diagram at large internuclear separations R, follows a potential curve until a minimum distance is reached and then recedes the same way it came in.

- *Diabatic collisions*: In regions of strong curvature of potential lines (radial coupling) or at very small internuclear separations (rotational coupling) strong couplings become important. The regions, in which two potential curves with the same m-value, e.g., σ and π, come very close to each other, are denoted as *avoided crossings* or *pseudocrossings* indicated by boxes in Fig. 2.5. In these regions, the electron may jump from one potential curve to another, so that on the outgoing branch of the collision, the electron ends up in a different state of one of the separated atoms. This is the quasimolecular excitation mechanism as described qualitatively before.

When examining the avoided crossings in a complex correlation diagram, one may predict, where transitions between potential energy curves occur and hence, which final states are populated. This gives rise to diabatic correlation rules, originally proposed by Barat and Lichten [21], which were based on qualitative considerations and which have been widely used for interpreting experimental results. Later, in extensive calculations by Eichler et al. [15], a large number of realistic correlation diagrams for complex systems have been constructed, which clearly exhibit a large series of avoided crossings. An example is given in Fig. 2.5. In this way, a new correlation rule has been established [15].

The experimental interest in very heavy quasimolecular collision systems, in particular in view of the possible formation of superheavy quasimolecules, has motivated the application of relativistic self-consistent methods. Several MO correlation diagrams have been calculated within the Dirac-Fock-Slater approximation [22, 23, 24].

In general, it is always possible that during the collision electromagnetic transitions occur. Such transitions have been identified in inner-shell excitations and give rise to *quasimolecular x-rays*, which will be discussed in the following.

2.5 Molecular orbital (MO) x-rays

The quasimolecular excitation mechanism has been confirmed by numerous studies of inner-shell excitations in slow collisions induced by radial and rotational couplings, see Sec. 2.8. However, it has been desirable to perform a direct spectroscopy of molecular orbitals as a function of the internuclear separation R by analyzing the MO x-rays emitted by electrons de-exciting into the inner-shell vacancies during the collision. This goal has not been reached in a clear-cut way as long as the inner-shell vacancies were created at an unspecified time during the same collision process.

The situation changed, when it became experimentally possible to strip the projectiles almost completely at high energies and subsequently decelerate them to velocities suitable for quasimolecular excitation. For example,

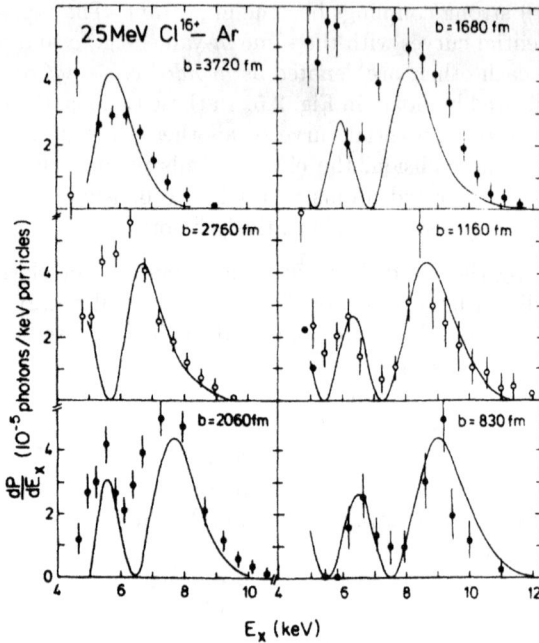

Figure 2.6: X-ray emission probabilities as a function of x-ray energy at different impact parameters for the beam energy of 2.5 MeV in Cl^{16+} + Ar collisions. The lines represent theoretical results. From [25].

by decelerating hydrogen-like Cl^{16+} projectiles, it is possible to introduce K-shell vacancies into the collision from the outset [25]. Hence, in Cl^{16+} + Ar collisions, the $1s\sigma$ vacancies can be filled from the $2p\pi$ level at any time during the collision, both on the incoming and on the outgoing branch of the collision trajectory.[†] This means that MO x-rays with the same energy E_x can be emitted at two different collision times corresponding to the same R. The resulting transition amplitudes interfere with each other and produce an interference structure in the x-ray spectrum.

Since $R(t)$ depends sensitively on the impact parameter b, it is experimentally indispensable to determine b by converting the scattering angle to the impact parameter, either by using the Coulomb deflection function

---

[†]The levels $1s\sigma$ and $2p\pi$ correspond to the levels 1σ and 1π, respectively in Figs. 2.4 and 2.5.

(1.13) or a screened version of it. The experimental results are shown in Fig. 2.6 [25] for Cl^{16+} + Ar collisions at 2.5 MeV for various impact parameters. The theoretical curves have been obtained from two different approaches, which are in good agreement with each other [25]. These results have produced the first experimental determination of $2p\pi$–$1s\sigma$ transition energies as a function of the internuclear distance R. In this way, a unique confirmation of the quasimolecular excitation mechanism has been established.

2.6 Two levels: Landau-Zener approximation

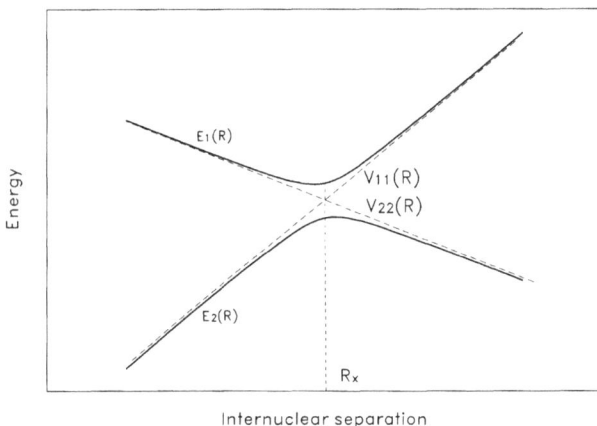

Figure 2.7: Schematic Landau-Zener diagram of adiabatic $(E_1(R), E_2(R))$ and diabatic $(V_{11}(R), V_{22}(R))$ potentials in the vicinity of the crossing at $R = R_x$.

If there is a clear pseudocrossing of two levels well separated from other levels, one may adopt a two-state approximation for estimating transition probabilities. With certain simplifying assumptions, one derives the famous *Landau-Zener approximation* (for an analysis and extensions, see, e.g., [2, 26]) for the transition probability

$$P_{LZ} = \exp\left(-\frac{2\pi V_{12}^2}{|V_{11}'(R) - V_{22}'(R)|\, v}\right),\qquad (2.26)$$

where V_{11}' and V_{22}' denote the slopes of the two *diabatic* potential curves near the pseudocrossing and V_{12} the diabatic coupling potential, which is estimated from the adiabatic energies $E_i(R)$ as $V_{12} = \frac{1}{2}\Delta E = \frac{1}{2}|E_1(R_\mathrm{x}) - E_2(R_\mathrm{x})|$. Furthermore, v is the velocity at the crossing point R_x. For a schematic diagram, see Fig. 2.7. It follows from Eq. (2.26) that the transition probability increases with increasing differences in the diabatic slopes V_{11} and V_{22}, with increasing velocity at the crossing point, and with a decreasing energy gap between the adiabatic potential curves. Formula (2.26) represents a widely used parametrization, however, often it is of a limited value. For a proper, but also numerically expensive treatment of the collision dynamics, one has to resort to the coupled-channel approach introduced in Sec. 2.1. An outline for molecular excitations is given in Secs. 2.8 and 2.9.

2.7   Two levels: Stückelberg oscillations

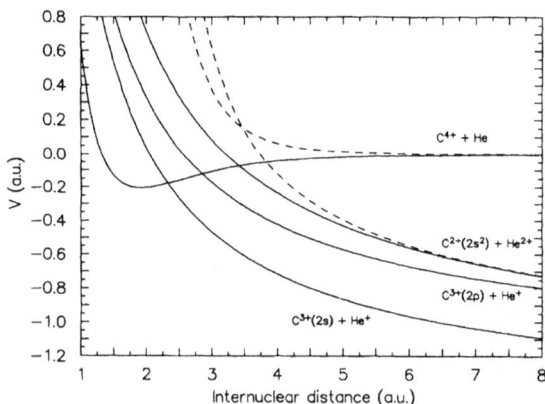

Figure 2.8: Diabatic potential energy curves from the models of Danared and Bárány [28] (dashed lines) and of Barat et al. [29] (solid lines). From [30].

Let us consider an atomic process, in which the system enters along the adiabatic curve $E_2(R)$, see Fig. 2.7, (diabatic curve $V_{22}(R)$) and leaves the reaction zone on the curve $E_1(R)$ (diabatic curve $V_{11}(R)$), or *vice versa*. According to energy and impact parameter, the collision is assumed to reach a minimum distance of $R_\mathrm{min} < R_\mathrm{x}$. Transitions between these two curves can happen on the incoming or on the outgoing branch of the reaction. Both paths leading to the same scattering angle θ are indistinguishable and hence will interfere, similarly as in an optical two-slit experiment. This has been first recognized by Stückelberg in 1932 [27]. The resulting interference pattern is denoted as Stückelberg oscillation. After integration over impact parameters, these oscillations are often smoothed out and do not become visible in experiments, in particular at high energies.

As a well-studied example, we consider the two-electron transfer [28, 29, 30] in the reaction

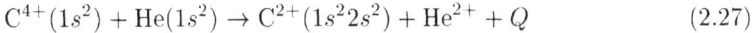

$$C^{4+}(1s^2) + He(1s^2) \to C^{2+}(1s^2 2s^2) + He^{2+} + Q \tag{2.27}$$

for low impact energies starting from 400 eV, where Q is the energy release. In this reaction, below 10 keV collision energy, double-electron transfer is much more likely to occur than single-electron transfer. Danared and Bárány [28] treated the system as a two-level system. Introducing the known polarizability α of the He atom, $q = 4$ and an effective repulsive potential C_N/R^n, $n > 4$, they parametrized the diabatic potentials as

$$V_{11} = -\frac{\alpha}{2}\frac{q^2}{R^4} + \frac{C_n}{R^n} \quad \text{and} \quad V_{22} = \frac{2(q-2)}{R} + \frac{C_n}{R^n} - Q, \tag{2.28}$$

where the parameters C_n and n as well as an interaction term V_{12} have been adjusted to pre-existing experiments. In a simple description, they used the classical deflection function (1.12) and took into account the change in the reaction potential at the crossing point in the incoming or in the outgoing part of the collision. For each scattering angle θ, two impact parameters b are contributing. Hence the classical cross section (1.14) at this angle is composed of two contributions. The probability p_i for a transition between the curves is given by the Landau-Zener expression (2.26). The scattering amplitude for a double passage within this model is given by the square root of the transition probability multiplied with a phase factor as

$$\rho_i = \left(p_i(1-p_i)\frac{d\sigma_i^{\text{class}}}{d\theta}\right)^{1/2} \exp(\text{i}A_i), \tag{2.29}$$

where $A_i = \int V_i dt$ is the classical action along the path i. The differential cross section then becomes

$$\frac{d\sigma}{d\theta} = |\rho_1 + \rho_2|^2. \tag{2.30}$$

Clearly, interferences and hence oscillations in the cross section arise from the phase difference $\phi = A_1 - A_2$ along the two paths. This is the simplest way to visualize the origin of the oscillations, but there are, of course, more refined treatments [28]. Barat et al. [29] considerd higher

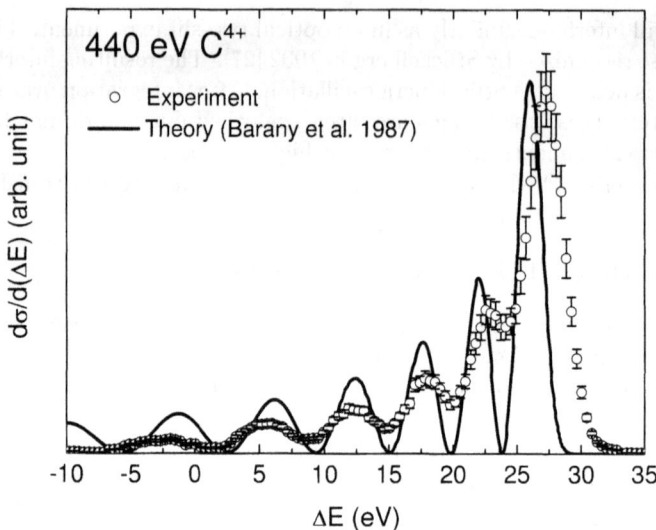

Figure 2.9: Energy gain spectrum of two-electron capture by C^{4+} from He at 440 eV. Circles: experiment, solid line: theory [32]. From [31].

energies and treated the coupling among four levels which, however, provide only an indirect coupling via $C^{3+}(2s) + He^+$ between initial and final states.

In Fig. 2.8, we show the potential energy curves used in [28] and [29]. While incoming and outgoing branches are identical, the details are quite different. Nevertheless, their calculations essentially reproduce the important features of their experimental results. Instead of using classical or semi-classical calculations, Keller et al. [30] performed two-channel and four-channel calculations, respectively, with a quantum mechanical treatment of the nuclear motion, adopting the potential curves of [28] and [29].

Stückelberg oscillations may be studied as a function of the scattering angle or, alternatively, as a function of the energy gain ΔE, that is the energy difference of the projectile before and after the collision, which is a function of the Q-value and the scattering angle. Both dependencies can also be measured simultaneously and presented in a contour diagram. Fig. 2.9 shows experimental results by Hoshino et al. [31] obtained as the projection of the two-dimensional plot on the energy axis and, for comparison, a theoretical curve derived from Bárány et al. [32]. One obtains a very good agreement between calculated and measured oscillations. More detailed results are contained in the contour plot as a function of both the scattering angle and the energy gain [33]. One then can show that the final state is $C^{2+}(1s^2\ 2s^2\ {}^1S)$ with a $Q=33.4$ eV.

2.8 Molecular orbitals: Dynamical calculations

Leaving the instructive two-level systems, we resume the general case. Similarly as in Sec. 2.4, following [11, 18], we consider a slow collision, in which a quasimolecule is transiently formed with a time-varying internuclear separation $R(t)$ and the corresponding interaction. Within the coupled-channel approach, Sec. 2.1, the set of basis states to be chosen, consists of the orthogonal and complete set of static molecular orbitals (MO), $\chi_k(\mathbf{r}, R)$, defined as eigenstates of the single-particle Hamiltonian (2.25) and of the operator for the angular momentum component along the internuclear line with eigenvalue m. They satisfy

$$H\left(\mathbf{r}, \mathbf{R}(t)\right)\chi_k\left(\mathbf{r}, \mathbf{R}(t)\right) = \epsilon_k\left(\mathbf{R}(t)\right)\chi_k\left(\mathbf{r}, \mathbf{R}(t)\right) \tag{2.31}$$

and depend parametrically on $\mathbf{R}(t)$. In order to investigate the time development of the electronic wavefunction $\Psi(t)$, we expand Ψ, separately for each value of m, in terms of molecular states as

$$\Psi(t) = \sum_k a_k(t)\, \chi_k\left(\mathbf{r}, \mathbf{R}(t)\right)\, \exp\left[-\mathrm{i}\int^t \epsilon_k\left(\mathbf{R}(t')\right)\mathrm{d}t'\right], \tag{2.32}$$

where the potential energies $\epsilon_k\left(\mathbf{R}(t)\right)$, defined in Eq. (2.31), are identified with the curves in the correlation diagrams, for example, Figs. 2.4 and 2.5. The states χ_k are fixed to the rotating internuclear axis. For the time being, we ignore the translation factors and include them later. Next, we insert the expansion (2.32) into the time-dependent Schrödinger equation (2.7), multiply with the complex conjugate of χ_j times the corresponding exponential and subsequently integrate over the electronic coordinate. Within the semiclassical approximation, the time derivative in Eq. (2.7) has a simple meaning:

$$\frac{\partial}{\partial t} \to v_R\frac{\partial}{\partial R} + \dot{\theta}\frac{\partial}{\partial \theta} \to v_R\frac{\partial}{\partial R} + \mathrm{i}\dot{\theta}L_y, \tag{2.33}$$

where $v_R = \dot{R}$ is the velocity along the internuclear line and L_y is the angular momentum operator with respect to the axis of rotation of the quasimolecule perpendicular to the x, z scattering plane. The operator L_y couples only molecular states whose angular momenta around the internuclear line differ by one unit. For orthogonal and normalized molecular eigenstates χ_k, the diagonal matrix elements of the operator $\partial/\partial R$ vanish as well, while the overlap matrix $\mathbb{N}$ reduces to the unit matrix. The coupled equations corresponding to Eq. (2.8) then take the form

$$i\dot{a}_j(t) \;=\; -\epsilon_j a_j(t)$$

$$+ \; i\sum_{k\neq j} a_k(t) v_R \left\langle \chi_j \left| \frac{\partial}{\partial R} \right| \chi_k \right\rangle \exp\left[-i\int^t (\epsilon_k - \epsilon_j)dt'\right]$$

$$+ \; i\sum_{k\neq j} a_k(t)\dot{\theta} \left\langle \chi_j \left| iL_y \right| \chi_k \right\rangle \exp\left[-i\int^t (\epsilon_k - \epsilon_j)dt'\right], (2.34)$$

where the coupling terms $v_R \langle \chi_j|\partial/\partial R|\chi_k \rangle$ and $\dot{\theta}\langle \chi_j|iL_y|\chi_k \rangle$ are denoted as the radial and the rotational (Coriolis) couplings, respectively. If one uses an expansion in terms of states that are *not* molecular (adiabatic) eigenstates, e.g., "diabatic" states, on has, in addition, potential couplings, because the Hamiltonian is not diagonalized in the diabatic representation.

The radial couplings become important near pseudocrossings, because the states and hence the matrix elements change rapidly in this vicinity. Since for a given straight-line trajectory $\dot{\theta} \approx vb/R^2$, the rotational coupling is dominant at small internuclear separations.

The wavefunctions χ_k used so far, do not obey proper boundary conditions for moving atomic orbitals. If the origin is taken at the target nucleus, the amplitudes that correlate to transfer states do *not* constitute transition amplitudes to the corresponding travelling atomic transfer states.

Only the total transfer probability can be defined unambiguously as everything that is left over from non-transfer states, that is

$$P_{\text{trans}} \;=\; \sum_{\substack{\text{all transfer states}}} |a_k(\infty)|^2 = 1 - \sum_{\substack{\text{all non-transfer states}}} |a_k(\infty)|^2.$$

$$(2.35)$$

In order to ensure proper travelling waves, one has to introduce an electron translation factor (ETF), see Sec. 2.2, so that

$$\psi_k(\mathbf{r}, R) = \chi_k(\mathbf{r}, R) f(\mathbf{r}, R) \tag{2.36}$$

with

$$f(\mathbf{r}, R) \rightarrow \begin{cases} 1 & \text{for} \quad R \rightarrow 0 \\ e^{i\mathbf{v}\cdot\mathbf{r}} & \text{for} \quad R \rightarrow \infty. \end{cases} \tag{2.37}$$

The choice of the "switching factor" $f(\mathbf{r}, R)$ interpolating between the two limits in the intermediate regime at finite values of R is ambiguous, for a discussion see, e.g., [11].

In summary, aside from this problem, the evaluation of the coupled equations (2.8) or (2.34) proceeds by considering the following points.

- The MO are defined with respect to the internuclear line, which rotates while the projectile passes the target atom.

- Because there is just a single molecular Hamiltonian (not two separated atomic ones), all states are orthogonal and normalized during the collision, so that the overlap matrix is $N_{jk} = \delta_{jk}$.

- Two types of coupling matrix elements occur:
  Rotational coupling,
  mediated by the matrix elements $\langle \chi_k | iL_y | \chi_j \rangle$ (where L_y is the angular momentum operator perpendicular to the scattering plane) connecting different m-states, e.g., $\sigma - \pi$. Because of the speed of rotation, these couplings are important for small values of R.

- *Radial coupling,*
  mediated by the matrix elements $\langle \chi_k | \partial/\partial R | \chi_j \rangle$ connecting states with the same m-value, e.g., $\sigma - \sigma$ or $\pi - \pi$. These couplings are important at avoided crossings.

The calculations are mainly impaired by the ambiguity of finding appropriate electron translation factors.

2.9 Two-center atomic expansions

In the intermediate velocity range, $v \approx v_e$, and for distant collisions, one has to give up the molecular (MO) description and, instead, expand in terms of atomic (AO) wavefunctions centered around target and projectile nuclei. If one has a sufficiently large basis set, one may also represent molecular orbitals as linear combinations of atomic orbitals (LCAO), i.e., carry the treatment to $v \ll v_e$. Moreover, one may place a third center in the middle between the nuclei in order to achieve an improved representation of MO [34]. Of course, in principle, each adiabatic Hamiltonian generates a complete set of eigenstates. Consequently, if one has two or three centers, the combined set is *overcomplete*, which may give rise to problems. However, in practice, this never occurs, because each of the sets is heavily truncated in numerical calculations, in particular the continua are usually cut off.

Explicitly, and taking into account the translation factors (2.19), (now denoting target states by unprimed subscripts and projectile states by primed labels) the expansion (2.12) is written as

$$\Psi(\mathbf{r}, t) = \sum_k a_k(t) \phi_k(\mathbf{r_T}) e^{-i\epsilon_k t} + \sum_{k'} a_{k'}(t) \phi_{k'}(\mathbf{r_P}) e^{i\mathbf{v}\cdot\mathbf{r} - \frac{i}{2}v^2 t} e^{-i\epsilon_{k'} t}$$

$$(2.38)$$

with

$$H_T \phi_k(\mathbf{r}_T) = \epsilon_k \phi_k(\mathbf{r}_T), \qquad \mathbf{r}_T = \mathbf{r}$$
$$H_P \phi_{k'}(\mathbf{r}_P) = \epsilon_{k'} \phi_{k'}(\mathbf{r}_P) \tag{2.39}$$

being the unperturbed Schrödinger equations of target and projectile in their own refence frame, see Eq. (2.14).

The overlap matrix elements are

$$N_{jk}(t) = \delta_{jk}$$
$$N_{jk'}(t) = \langle \phi_j | e^{i\mathbf{v}\cdot\mathbf{r}} | \phi_{k'} \rangle \, e^{-\frac{i}{2}v^2 t} e^{-i(\epsilon_{k'}-\epsilon_j)t}$$
$$N_{j'k}(t) = \langle \phi_{j'} | e^{-i\mathbf{v}\cdot\mathbf{r}} | \phi_k \rangle \, e^{\frac{i}{2}v^2 t} e^{-i(\epsilon_k-\epsilon_{j'})t}$$
$$N_{j'k'}(t) = \delta_{j'k'} \tag{2.40}$$

For straight-line trajectories (for curved trajectories, one has, in principle, an additional acceleration term $\dot{\mathbf{v}} \cdot \mathbf{r}$), the coupling matrix elements are

$$M_{jk}(t) = \langle \phi_j | \epsilon_k - H | \phi_k \rangle \, e^{-i(\epsilon_k-\epsilon_j)t}$$
$$M_{jk'}(t) = \langle \phi_j | e^{i\mathbf{v}\cdot\mathbf{r}}(\epsilon_{k'} - H) | \phi_{k'} \rangle \, e^{-\frac{i}{2}v^2 t} e^{-i(\epsilon_{k'}-\epsilon_j)t}$$
$$M_{j'k}(t) = \langle \phi_{j'} | e^{-i\mathbf{v}\cdot\mathbf{r}}(\epsilon_k - H) | \phi_k \rangle \, e^{\frac{i}{2}v^2 t} e^{-i(\epsilon_k-\epsilon_{j'})t}$$
$$M_{j'k'}(t) = \langle \phi_j | \epsilon_{k'} - H | \phi_{k'} \rangle \, e^{-i(\epsilon_{k'}-\epsilon_{j'})t}. \tag{2.41}$$

If the basis states are originally not atomic eigenstates (i.e., pseudostates, see Sec. 2.10), one may prediagonalize each basis set attached to one center in order to obtain eigenstates.

The coupled equations

$$\dot{\mathbf{a}} = -i\mathbb{N}^{-1}\mathbb{M}\mathbf{a} \qquad \text{with} \qquad \mathbb{N} \text{ and } \mathbb{M} = \begin{pmatrix} TT & TP \\ PT & PP \end{pmatrix}, \tag{2.42}$$

are solved as before. The matrix to the right, schematically indicates the subspaces of target and projectile. The initial conditions for the system being in the state k_0 are

$$a_{k_0}(-\infty) = 1$$
$$a_k(-\infty) = 0 \qquad \text{for} \qquad k \neq k_0$$
$$a_{k'}(-\infty) = 0. \tag{2.43}$$

The benchmark system for atomic collision models is the system H$^+$ + H. Since the process of transfer into the 1s state of the projectile is a resonant process, the total transfer cross section has already been well reproduced in the pioneering investigations with two-state MO and two-state

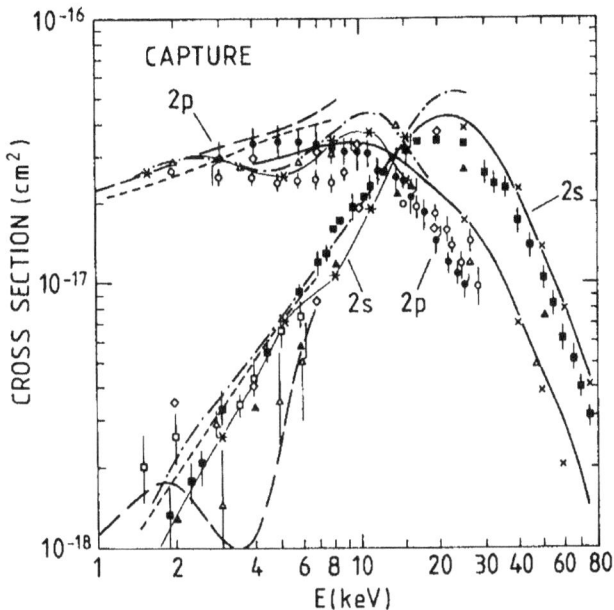

Figure 2.10: Cross sections for electron transfer into projectile 2s and 2p states in $H^+ + H$ collisions. Calculated results from basis-expansion studies: short-dashed and long-dashed lines: MO-basis calculations; dash-dotted and solid lines: AO expansion by Fritsch and Lin [35] and [36], respectively, the latter with a larger basis set including pseudostates. Various other calculations are indicated by symbols *without* error bars, while experimental data are given *with* error bars. For details and references, see [11].

AO expansions including translation factors. More recent investigations have concentrated on the evaluation of excitation and transfer processes to 2s and 2p states, which are, at low energies, one or two orders of magnitude weaker than the dominant 1s–1s transition. They provide a more sensitive test of the model description. As an example, we include a comparison between experimental data and the results of various calculations in Figs. 2.10 and 2.11 [11].

It is remarkable that two-center atomic-orbital expansions adopting translation factors and pseudostates can reproduce even sensitive cross sections over a wide range of collision energies, including low energies where MOs are expected to be formed transiently.

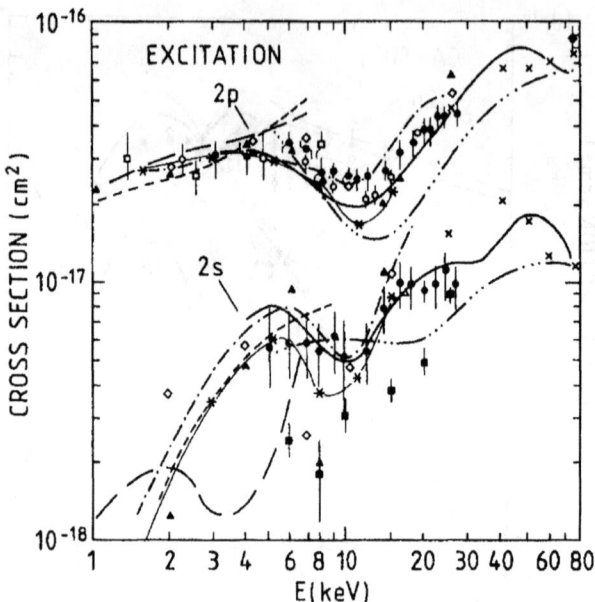

Figure 2.11: Cross sections for electron excitation into target 2s and 2p states in $H^+ + H$ collisions. See caption of Fig. 2.10. The triple-dot-dash curves represent results from AO expansions with an approximate inclusion of the remaining space of basis states. From [11].

The theoretical analysis can be extended to other light atoms and ions and has proven to be very helpful in the diagnosis of fusion plasmas by examining optical transition lines between excited states, which in some cases are inaccessible by direct experiments.

2.10   Mathematical appendix: Single-center basis functions

In Sections 2.8 and 2.9, we have assumed basis expansions in terms of molecular eigenstates derived by solving the static molecular eigenvalue problem or, alternatively, in terms of atomic eigenstates of each center defined by Eq. (2.14). However, sometimes, it is convenient to expand in terms of basis functions that are mathematically convenient rather than solutions of physical eigenvalue problems. Besides atomic eigenstates, we here consider pseudostates represented by Slater-type orbitals as well as two other types that have been widely applied, namely Sturmian and Gaussian basis states.

Hydrogenic eigenstates

For low-Z single-electron systems, it is suggestive to adopt exact nonrelativistic eigenfunctions of Eq. (2.14) as basis states for each center. For a given Z and with $k = \{n, l, m\}$, the hydrogenic functions are defined by

$$\phi_{nlm}(\mathbf{r}) = r^{-1} R_{nl}(r) Y_{lm}(\hat{\mathbf{r}}), \tag{2.44}$$

where $Y_{lm}(\hat{\mathbf{r}})$ is a spherical harmonic depending on the unit vector $\hat{\mathbf{r}} = (\theta, \varphi)$, while the normalized radial functions (in atomic units) [37] are

$$
\begin{aligned}
R_{nl}(r) &= \left(\frac{Z}{n} \frac{(n-l-1)!}{n[(n+l)!]^3} \right)^{1/2} e^{-x/2} x^{l+1} L_{n-l-1}^{2l+1}(x), \\
&= \frac{1}{(2l+1)!} \left(\frac{Z}{n} \frac{(n+l)!}{n(n-l-1)!} \right)^{1/2} e^{-x/2} x^{l+1} \\
&\qquad \times {}_1F_1(-n+l+1, 2l+2, x) \tag{2.45}
\end{aligned}
$$

with

$$x = 2Zr/n.$$

Here,

$$L_q^p(x) = \sum_{\mu=0}^{q} (-1)^\mu \binom{q+p}{q-\mu} \frac{x^\mu}{\mu!} \tag{2.46}$$

is an associated Laguerre polynomial and ${}_1F_1(a, b, c)$ the confluent hypergeometric function [38]. While the functions defined by (2.39) are directly the solutions for each atomic center of a two-center single-electron system at large separations (aside from a phase distortion, see Sec. 4.3), one obtains a better representation of *molecular* wavefunctions at smaller separations, if one augments the set by additional states, which are hydrogenic eigenstates of the charge $Z = Z_P + Z_T$ in the united-atom limit. In general, one may add hydrogenic eigenstates of arbitrary charges. Such states are called *pseudostates*. Clearly, pseudostates belonging to different charges are no longer orthogonal with respect to each other. This means that in Eq. (2.40) the Kronecker deltas in the first and last equation have to be replaced by the appropriate overlap matrix element in cases with different

Z. Pseudostates have overlap also with continuum states of the physical charge. Therefore, the inclusion of pseudostates gives a representation of part of the continuum. However, there is no clear-cut prescription, in which way pseudostates should be selected in order to speed up convergence.

An alternative way to construct discrete continuum states is by superposing adjacent continuum eigenstates of Eq. (2.14), which can be separated in parabolic coordinates [39], within a given energy interval (see Eq. (7.11) and Fig. 7.3, which are similarly valid for the nonrelativistic case). Introducing the asymptotic wave vector $\mathbf{k} = \mathbf{p}/\hbar = \mathbf{v}$ (in a.u.) of the electron and the Sommerfeld parameter $\nu = Z/v$, we can write the solution for *outgoing* spherical electron waves, to be used for *electron absorption* as

$$\phi_{\mathbf{k}}^{(+)}(\mathbf{r}) = e^{\frac{\pi}{2}\nu}\,\Gamma(1 - i\nu)\,e^{i\,\mathbf{k}\cdot\mathbf{r}}\,{}_1F_1\left[i\nu\,,1\,;i\,(kr - \mathbf{k}\cdot\mathbf{r})\right] \qquad (2.47)$$

and for *incoming* spherical waves, to be used for *electron emission*, as

$$\phi_{\mathbf{k}}^{(-)}(\mathbf{r}) = e^{\frac{\pi}{2}\nu}\,\Gamma(1 + i\nu)\,e^{i\,\mathbf{k}\cdot\mathbf{r}}\,{}_1F_1\left[-i\nu\,,1\,;-i\,(kr + \mathbf{k}\cdot\mathbf{r})\right]. \qquad (2.48)$$

These wavefunctions are normalized such that for $Z \to 0$ they merge into plane waves $\phi_{\mathbf{k}}(\mathbf{r}) \to \exp(i\,\mathbf{k}\cdot\mathbf{r})$. Instead of using the separation in parabolic coordinates, which yields the closed-form expressions (2.47) and (2.48), the Coulomb continuum functions can also be formulated in a partial-wave expansion, see Eq. (3.30).

Slater-type orbitals (STO)

Pseudostates can also be introduced by abandoming the requirement of hydrogenic eigenstates, that is by giving up the structure (2.45) of the radial functions altogether and by using instead simple radial functions denoted as Slater-type orbitals (STO) with the replacements in Eq. (2.44)

$$R_{nl}(r) \to \chi_i(r) = N_i\,r^{n_i}\exp(-\zeta_i r) \qquad (2.49)$$

with

$$N_i = \frac{(2\zeta_i)^{n_i+1/2}}{[\Gamma(2n_i + 1)]^{1/2}}.$$

Here, for each angular momentum l, one chooses a set of quantum numbers n_i and orbital exponents ζ_i, so that, in the first place, one obtains a good approximation to the atomic eigenenergy by diagonalizing the atomic Hamiltonian. This method is also applicable for calculating the eigenstates of a non-Coulombic effective potential (say a Hartree-Fock-Slater potential) generated by inner-shell electrons and acting on a single active electron. Furthermore, suitably chosen STO's may be added to improve convergence in a two-center expansion (2.38). Again, STO's are not orthogonal to each other. The drawback of this type of expansion is that there exists no systematic way to achieve good convergence. This defect is avoided with Sturmian functions to be discussed next.

Sturmian basis states

Sturmian functions have been introduced into atomic basis expansions by Rotenberg in 1962 [40] and brought into the presently used form by Gallaher and Wilets in 1968 [41]. They are not eigenfunctions of a physical system except for one of the set. Instead, they are generated by a differential equation closely related to the Schrödinger equation.

The eigenfunctions $\phi_k(\mathbf{r})$ in Eq. (2.39) are replaced by products of eigenfunctions of angular momentum and by radial functions $S_{nl}(r)$ in the form

$$\phi_k(\mathbf{r}) = r^{-1} S_{n_k l_k}(r) Y_{l_k m_k}(\hat{\mathbf{r}}), \tag{2.50}$$

defined in analogy to Eq. (2.44). The Sturmian functions $S_{nl}(r)$ for a hydrogenic system with charge Z are the solutions (in atomic units) of

$$\left[-\frac{1}{2} \frac{d^2}{dr^2} + \frac{l(l+1)}{2r^2} - \frac{\alpha_{nl} Z}{r} \right] S_{nl}(r) = E_l S_{nl}(r), \tag{2.51}$$

where E_l is a fixed parameter. The eigenvalue α_{nl} is an effective charge, adjusted so that $S_{nl}(0)$ is zero and $S_{nl}(r)$ decays exponentially as r tends to infinity. This can only be achieved for discrete values of α_{nl}. Equation (2.51) looks almost like the Schrödinger equation. The difference is that the binding energy is kept constant while the coupling strength is varied until the boundary conditions are met.

By introducing a scale transformation $\rho = \alpha_{nl} r$, we obtain from Eq. (2.51)

$$\left[-\frac{1}{2} \frac{d^2}{d\rho^2} + \frac{l(l+1)}{2\rho^2} - \frac{Z}{\rho} \right] S_{nl}(\rho/\alpha_{nl}) = \frac{E_l}{\alpha_{nl}^2} S_{nl}(\rho/\alpha_{nl}), \tag{2.52}$$

which shows that the functions $S_{nl}(\rho/\alpha_{nl})$ are solutions of a hydrogen-like problem, so that $S_{nl}(\rho/\alpha_{nl}) \propto R_{nl}(\rho)$ or $S_{nl}(r) \propto R_{nl}(\alpha_{nl} r)$. Here, the hydrogenic radial functions $R_{nl}(r)$ are defined in Eq. (2.45). For a given angular momentum l, the principal quantum number n takes the values $n = l + 1, l + 2, \cdots$ and the number $n_r = n - l - 1$ counts the number of radial nodes. From the normalization of the hydrogenic radial functions and the scale transformation, it follows that

$$S_{nl}(r) = \alpha_{nl}^{1/2} R_{nl}(\alpha_{nl} r) \tag{2.53}$$

with the normalization

$$\int_0^\infty [S_{nl}(r)]^2 dr = 1. \tag{2.54}$$

Assuming a negative value of the parameter E_l, the Sturmian functions (2.50) with the radial part (2.53) form an infinite, discrete, and complete set of states. However, they are not orthogonal in the usual sense. Instead, one derives a modified orthogonality condition by writing the Sturmian equation (2.51) for n and n', multiplying, respectively, with $S_{n'l}$ and S_{nl} from the left, integrating over the radial coordinate and, finally, subtracting the integrals. As a result, one obtains, independently of Z,

$$(\alpha_{nl} - \alpha_{n'l}) \int S_{nl}(r) \frac{1}{r} S_{n'l}(r) dr = 0, \tag{2.55}$$

which is equivalent to an "orthogonality relation" with the weight factor $1/r$. The normalization of $\int S_{nl} \frac{1}{r} S_{nl} dr$ follows from the corresponding scaled expectation value $\langle 1/r \rangle = Z/n^2$ for hydrogenic functions [39], so that

$$\int_0^\infty S_{n'l}(r) \frac{1}{r} S_{nl}(r) dr = \delta_{nn'} \frac{\alpha_{nl} Z}{n^2}. \tag{2.56}$$

At this point, it is convenient [41] to choose the energy parameter as

$$E_l = -\frac{Z^2}{2(l+1)^2}. \tag{2.57}$$

Now equating the effective energy E_l/α_{nl}^2 appearing on the right-hand side of Eq. (2.52) to the hydrogenic energy $-Z^2/2n^2$ and inserting E_l from Eq. (2.57), one finds

$$\alpha_{nl} = \frac{n}{l+1}, \tag{2.58}$$

independently of the charge Z. From the choice (2.57), it follows that for a given angular momentum l, the Sturmian state with the lowest possible quantum number $n = l + 1$ (that is, with no radial nodes) is identical to the corresponding hydrogenic state, namely 1s, 2p, 3d, $\cdots$.

Inserting Eq. (2.58) into Eq. (2.56), one obtains for the modified orthogonality relation

$$\int_0^\infty S_{n'l}(r) \frac{1}{r} S_{nl}(r) dr = \delta_{nn'} \frac{Z}{n(l+1)}. \tag{2.59}$$

Comparing with the hydrogenic value Z/n^2 and taking into account that $l + 1 \leq n$, one notices that the diagonal Sturmian $1/r$ matrix elements

(2.59) are larger than the hydrogenic analogues, and hence the Sturmian functions are more compact.

The Sturmian functions form an enumerable and complete infinite set of discrete square-integrable states. Unlike for the hydrogenic functions, there is no continuum. The Sturmian functions have been proven to be very powerful in representing the hydrogenic continuum. Owing to their unique construction, they lend themselves to systematic convergence studies.

Gaussian basis functions

Besides the use of Slater-type orbitals or Sturmian functions, there is another possibility for an expansion around each of the atomic centers in Eq. (2.38), namely by linear combinations of Gaussian functions (GTO) in the form

$$\phi_{nlm}(\mathbf{r}) = \sum_{\nu} C_{\nu}^{(nl)} e^{-\alpha_{\nu} r^2} r^l Y_{lm}(\hat{\mathbf{r}}), \qquad (2.60)$$

where the solid harmonics $r^l Y_{lm}(\hat{\mathbf{r}})$ can be written in Cartesian coordinates. Within this representation, see, e.g., [42], the single-center matrix elements of each atomic Hamiltonian are calculated analytically in Cartesian coordinates and, subsequently, the atomic eigenfunctions are obtained by numerical diagonalization.

Before performing the latter task, a choice has to be made for the Gaussian parameters α_{ν}. As has been dicussed in [43], it is economical for computations to choose the Gaussian width parameters in geometric progression, that is $\alpha_{\nu+1} = \rho \alpha_{\nu}$. However, if one wishes a good representation of continuum states, one may adopt a factor ρ which slowly varies with ν in such a way that large widths (small α_{ν}) are emphasized.

Chapter 3

High-energy collisions: Perturbation theory for direct reactions

In this Chapter, we discuss energetic direct reactions, that is, excitation and ionization, and defer rearrangement collisions to Chapter 4. While for slow collisions, see Chapter 2, the interaction time is so long that the projectile does not give rise to a *small* perturbation, this is different for fast collisions, when $v \gg v_e$. Then it is appropriate to resort to a perturbative treatment formulated in the impact parameter picture.

We start in Sec. 3.1 with presenting a general outline of a perturbative or Born expansion including all orders of perturbation theory and then specify to the first order, which is most commonly used. As an alternative approach, in Sec. 3.2, we discuss the Magnus expansion, which has the advantage of being unitary at each level of approximation but is more difficult to evaluate numerically, even in its first order. As a widely adopted general method, we derive the distorted-wave Born approximation (DWBA) in Sec. 3.3 and specialize to excitation and ionization. While the treatment of excitation is straightforward because it involves only bound states, ionization leads to continuum states, for which we present two different descriptions: (a) the partial-wave expansion for the continuum function in Sec. 3.4, (b) the continuum-distorted-wave-eikonal-initial-state (CDW-EIS) approximation in Sec. 3.5, which, to some extent, takes into account the distortion of the continuum wavefunction by both the target and the projectile but in an unsymmetric fashion.

As a general requirement for perturbation theory to be applicable, the action of the projectile exerted on the target during the collision along the path s should be small, i.e.,

$$\frac{1}{\hbar} \int_{-\infty}^{\infty} V_P \, dt \ll 1. \tag{3.1}$$

Specifically, assuming $V_P = (e^2 Z_P / r)$, we have (in a.u.)

$$\int_{-\infty}^{\infty} \frac{Z_P}{r} \, dt = \int_{-\infty}^{\infty} \frac{Z_P}{r} \frac{ds}{v} = \mathcal{O}\left(\frac{Z_P}{b} \frac{b}{v}\right), \tag{3.2}$$

so that Eq. (3.1) demands

$$\frac{Z_P}{v} \ll 1. \tag{3.3}$$

Hence Z_P/v is the decisive parameter for judging the strength or weakness of a (nonrelativistic) perturbation.

3.1   The Born expansion

Starting from our exact equation (2.10), see also [2], p. 249, the differential equation can be written in a compact form as

$$i \dot{\mathbf{a}} = \mathbb{N}^{-1} \, \mathbb{M} \mathbf{a} \equiv \mathbb{V} \mathbf{a}, \tag{3.4}$$

where $\mathbf{a}(t)$ is the column vector composed of the time-dependent amplitudes $a_k(t)$, while $\mathbb{N}$ and $\mathbb{M}$ are the overlap and coupling matrices, respectively. With the initial condition for the initial state ψ_i with the set of quantum numbers i

$$\lim_{t \to -\infty} \mathbf{a}(t) = \mathbf{a}^{(i)}, \quad \text{with} \quad a_j^{(i)} = \delta_{ij}, \tag{3.5}$$

the solution for this coupled system of equations satisfies the integral equation

$$\mathbf{a}(t) = \mathbf{a}^{(i)} - i \int_{-\infty}^{t} \mathbb{V}(t_1) \mathbf{a}(t_1) \, dt_1. \tag{3.6}$$

By iterating, we derive the Born expansion

$$\begin{aligned}
\mathbf{a}(t) \;=\; \mathbf{a}^{(i)} \Bigg[& 1 - i \int_{-\infty}^{t} \mathbb{V}(t_1) \, dt_1 \\
& + (-i)^2 \int_{-\infty}^{t} dt_1 \int_{-\infty}^{t_1} dt_2 \mathbb{V}(t_1) \mathbb{V}(t_2) + \cdots \Bigg],
\end{aligned} \tag{3.7}$$

where **1** denotes the unit matrix. If, in the last term, one extends the t_2-integration also to $t_2 > t_1$, so that it is unrestricted, one has to compensate this by a factor $1/2$ and, correspondingly, by $1/n!$ for the nth higher-order term. This leads to the Dyson expansion [44]

$$\mathbf{a}(t) = \mathbf{a}^{(i)} \left[\mathbf{1} + \sum_{n=1}^{\infty} \frac{(-i)^n}{\hbar^n n!} \int dt_1 \cdots \int dt_n \, \hat{T} \left[\mathbb{V}(t_1) \cdots \mathbb{V}(t_n) \right] \right]. \quad (3.8)$$

Here, we have introduced the Dyson time-ordering operator $\hat{T}$, which ensures that the time arguments decrease from left to right, so that

$$\hat{T} \left[X(t_i) \, X(t_j) \, X(t_k) \cdots \right] \quad = \quad X(t_1) \, X(t_2) \cdots X(t_n)$$
$$\text{with} \quad t_1 > t_2 \cdots > t_n. \quad (3.9)$$

In Eq. (3.8), we also have temporarily reintroduced the powers of $\hbar$ in order to indicate that the expansion is in terms of action integrals divided by $\hbar$ as in Eq. (3.1). Formally, the column vector of transition amplidudes can now be written as

$$\mathbf{a}(t) = \hat{T} \exp \left[-\frac{i}{\hbar} \int_{-\infty}^{t} \mathbb{V}(t') dt' \right] \mathbf{a}^{(i)}(t). \quad (3.10)$$

Assuming the validity of the condition (3.1) and the orthogonality of the states involved, so that $\mathbb{N} = \mathbf{1}$, we obtain the first-order transition amplitude (first-order perturbation theory) directly from Eq. (3.6) in the form

$$a_{fi}^{(1)} = a(\infty) = \delta_{if} - i \int_{-\infty}^{\infty} \langle \psi_f | V(\mathbf{r}, t') | \psi_i \rangle \, dt'. \quad (3.11)$$

We note that couplings of the initial and final states with other (intermediate) states are completely disregarded. Equation (3.11) is the basis of most perturbative approaches. In particular, the (first-order) Born approximation is obtained by representing the continuum state by an undistorted plane wave.

In all cases, the transition probability $P(v, b)$ for a given impact parameter b is obtained from Eq. (2.5) and the total cross section from (2.6). Clearly, for small impact parameters, the perturbation may become very strong, so that unitarity is violated by $P(v, b) > 1$ and first-order perturbation theory breaks down.

3.2   The Magnus expansion

From the Dyson representation (3.10), one may immediately derive another expansion, the Magnus expansion [45], which has been analyzed by

Pechukas and Light [46] and applied, e.g., to Coulomb excitation of nuclei [47] and to ionization [48]. The Magnus expansion rewrites the exponent of the expression (3.10) in terms of a sum involving nested commutators with an increasing number of potentials. The exact transition amplitude is written as

$$
\mathbf{a}(t) \;=\; \mathbf{a}^{(i)} \left\{ \exp\left[-\frac{i}{\hbar} \int_{-\infty}^{\infty} \mathbb{V}(t)\,dt \right.\right.
$$
$$
\left.\left. +\frac{1}{2}\left(-\frac{i}{\hbar}\right)^2 \int_{-\infty}^{\infty} dt \int_{-\infty}^{t} dt' \, [\mathbb{V}(t), \mathbb{V}(t')] + \cdots \right] \right\}.
$$

(3.12)

This expression can be considered as a continuous generalization of the Baker-Hausdorff theorem, see e.g., [46, 47]. Let $\mathbb{A}$ and $\mathbb{B}$ be two noncommuting matrices, then

$$
\exp\{\mathbb{A}\}\exp\{\mathbb{B}\} \;=\; \exp\{\mathbb{A} + \mathbb{B} + \tfrac{1}{2}[\mathbb{A}, \mathbb{B}]
$$
$$
+\tfrac{1}{12}[\mathbb{A}, [\mathbb{A}, \mathbb{B}]] + \tfrac{1}{12}[\mathbb{B}, [\mathbb{B}, \mathbb{A}]] + \cdots\}.
$$

(3.13)

In contrast to the perturbation series (3.8), the Magnus expansion leads to transition operators that are unitary at each level of approximation.

Now suppose that the collision is so fast that the time structure in the exponent of (3.12), that is, the sequence of events, plays no role. This is the case, if the collision time $\tau_{\text{coll}} \approx a_{\text{K}}/v = 1/(Z_{\text{T}}v)$ for acting on the K-shell is short compared to the K-shell orbiting time $\tau_{\text{orb}} = \omega_{\text{orb}}^{-1} \approx (Z_{\text{T}}^2/2)^{-1}$, that is if

$$
\frac{\tau_{\text{coll}}}{\tau_{\text{orb}}} \approx \frac{Z_{\text{T}}}{v} \ll 1.
$$

(3.14)

Note that the ratio Z_{T}/v and *not* Z_{P}/v is the important parameter. If the time structure is irrelevant, the commutators at different times vanish (to a good approximation), and we obtain the *first-order Magnus* or *sudden approximation* [47, 48] defined by

$$
a_{\text{fi}} = \left\langle \psi_{\text{f}} \left| \exp\left[-\frac{i}{\hbar} \int_{-\infty}^{\infty} V(\mathbf{r}, t)\,dt \right] \right| \psi_{\text{i}} \right\rangle,
$$

(3.15)

independently of the projectile charge Z_{P}, which governs perturbation theory. This approach is expected to be useful for strong, very fast interactions. Within perturbation theory, (3.11), unitarity is violated for strong projectile potentials and for small impact parameters. This is not the case for the first Magnus approximation. However, because of the rapid oscillations in the integrand of the matrix element, it is difficult to evaluate numerically. This may be the reason, why the approach has not been applied very frequently. By expanding the exponent of Eq. (3.15) to first order, the familiar first-order perturbation theory (3.11) is retrieved.

3.3 The distorted-wave Born approximation (DWBA)

A widely adopted approximation consists in replacing the unperturbed plane wave of the Born approximation, see Sec. 3.1, by distorted waves, in which some part of the perturbation is already taken into account. This procedure is expected to lead to a faster convergence of the perturbation series and to a better estimate for the first-order term, because the remaining perturbation becomes weaker. There are many possible ways to do this. The physical model then consists in making a good guess for the potential that produces a realistic distortion from the outset.

In essence, the basic idea is to include part of the interaction exerted by the distant nucleus on the active electron in the definition of the "initial channel" or, correspondingly, in the definition of the "final channel". The resulting wavefunctions are called "distorted waves". Of course, for asymptotic internuclear separations, they reduce to the relevant atomic eigenstates[†]. Subsequently, the scattering problems for the initial and final channels are solved exactly.

The starting point of the derivation is an exact representation of the transition amplitude as a projection of the asymptotic initial or final states on the exact outgoing- or incoming- wave solutions, respectively, in analogy to Eq. (2.21), see, e.g., [2, 3, 49]

$$a_{\mathrm{fi}}(b) = \lim_{t \to \infty} \left\langle \chi_{\mathrm{f}}^{-}(t) \,\middle|\, \psi_{\mathrm{i}}^{+}(t) \right\rangle = \lim_{t \to -\infty} \left\langle \psi_{\mathrm{f}}^{-}(t) \,\middle|\, \chi_{\mathrm{i}}^{+}(t) \right\rangle. \qquad (3.16)$$

Here, ψ_{i}^{+} is the exact outgoing solution specified by the initial state, and ψ_{f}^{-} is the exact incoming solution specified by the final state. Both of them can be used to deduce the transition amplitude by projecting on the final or initial distorted (or undistorted) states χ_{f}^{-}, χ_{i}^{+}, which asymptotically merge into the eigenstates of the corresponding atomic Hamiltonian. In other words, the transition amplitude asks, how much of the final state is asymptotically $(R \to \infty)$ contained in the exact outgoing solution having evolved from the initial state or how much of the initial state is asymptotically contained in the exact incoming solution evolving to the final

---

[†]For the special case of a Coulomb potential with infinite range, particular care has to be taken, see Sec. 4.3

state. This is the same reasoning as in the coupled-channel description, see
Sec. 2.1. On the other hand, the final distorted wave is not contained in
the exact initial solution at time $t \to -\infty$ and the initial distorted wave
not in the exact final solution at $t \to \infty$, except for elastic collisions, that
is

$$\lim_{t \to -\infty} \langle \chi_f^-(t) \mid \psi_i^+(t) \rangle = 0$$

$$\lim_{t \to \infty} \langle \psi_f^-(t) \mid \chi_i^+(t) \rangle = 0. \tag{3.17}$$

With the overlap terms vanishing at the lower or upper limit, respectively,
we may write for the first version in Eq. (3.16), denoted as the *post* amplitude,

$$
\begin{aligned}
a_{fi}(b) &= \lim_{t \to \infty} \langle \chi_f^- | \psi_i^+ \rangle \\
&= \int_{-\infty}^{\infty} dt\, \frac{\partial}{\partial t} \langle \chi_f^-(t) \mid \psi_i^+(t) \rangle \\
&= \int_{-\infty}^{\infty} dt \left\langle \chi_f^{-*}(t)\, \frac{\partial \psi_i^+(t)}{\partial t} + \frac{\partial \chi_f^{-*}(t)}{\partial t}\, \psi_i^+(t) \right\rangle \\
&= -i \int_{-\infty}^{\infty} \left(\left\langle \chi_f^-(t) \left| i \frac{\partial}{\partial t} \psi_i^+(t) \right\rangle - \left\langle i \frac{\partial}{\partial t} \chi_f^-(t) \right| \psi_i^+(t) \right\rangle \right).
\end{aligned}
\tag{3.18}
$$

Note that, in contrast to Eq. (3.16), the space integral has to be taken at
all times, not only at $t \to \pm\infty$, so that the distortion at finite internuclear separation matters and hence different assumptions for the distorted
waves lead to different integrands, although, by construction, the resulting transition amplitudes are identical unless further approximations are
introduced.

Now eliminating from (3.18) the time derivative acting on $\psi_i^+(t)$ with
the aid of the exact time-dependent Schrödinger equation, we obtain the
transition amplitude in the *post form* as

$$a_{fi}(b) = -i \int_{-\infty}^{\infty} dt \left\langle \left(H - i \frac{\partial}{\partial t} \right) \chi_f^-(t) \mid \psi_i^+(t) \right\rangle. \tag{3.19}$$

Alternatively, we derive the *prior form* as

$$a_{fi}(b) = -i \int_{-\infty}^{\infty} dt \left\langle \psi_f^-(t) \left| H - i \frac{\partial}{\partial t} \right| \chi_i^+(t) \right\rangle. \tag{3.20}$$

While both expressions (3.19) and (3.20) are identical and exact, the
DWBA approximation can be derived by substituting the distorted waves

for the exact solutions, that is,

$$\psi_i^+ (t) \longrightarrow \chi_i^+ (t)$$
$$\psi_f^- (t) \longrightarrow \chi_f^- (t). \tag{3.21}$$

If the distorted waves $\chi_i^+(t)$ and $\chi_f^-(t)$ are solutions of the time-dependent Schrödinger equation for an approximate model Hamiltonian H_{approx}, the replacement (3.21) leads for both Eqs. (3.19) and (3.20) to the DWBA amplitude

$$a_{fi}^{DWBA}(b) = -i \int_{-\infty}^{\infty} dt \left\langle \chi_f^- (t) \mid H - H_{approx} \mid \chi_i^+ (t) \right\rangle. \tag{3.22}$$

It should be noted that the identity between the post and the prior form is true for direct reactions (excitation and ionization) but *not in general* for rearrangement collisions because the approximate Hamiltonians for initial and final states are usually different. The physical interpretation of Eq. (3.22) is the following. Part of the exact Hamiltonian H is already included in the approximate distorted wavefunctions $\chi_i^+(t)$ and $\chi_f^-(t)$. Therefore, the perturbation is given by the *residual interaction* $H - H_{approx}$ that is *not* included in the approximate wavefunctions. For example, if one adopts the first-order Born approximation, that is, undistorted waves, only the kinetic-energy term cancels out in $H - H_{approx}$, so that the full potential has to be used as a perturbation and one will expect less realistic results.

The total cross section for excitation or ionization by a fast bare ion is given by (see Eq. (2.6))

$$\sigma_{fi}^{DWBA} = 2\pi \int_0^{\infty} \mid a_{fi}^{DWBA} \mid^2 b\, db \qquad \text{for excitation} \tag{3.23}$$

and

$$\frac{d\sigma_{fi}^{DWBA}}{dE_f} = 2\pi \int_0^{\infty} \mid a_{fi}^{DWBA} \mid^2 b\, db \qquad \text{for ionization.} \tag{3.24}$$

3.4 Excitation and ionization: Partial-wave expansion

For high-energy projectiles, excitation of an electron from one bound state to another and ionization of a bound electron into the continuum are the dominamt processes, see, e.g., [50]. Adopting the impact parameter picture, the total cross section for excitation or ionization by a fast bare ion is given by Eqs. (3.23) and (3.24), respectively. Within the Born approximation, that is the first-order perturbation theory for undistorted initial and final states $\phi_{i,f}(\mathbf{r})$, the transition amplitude (3.11) caused by a bare projectile with charge number Z_P has the explicit form

$$a_{fi} = i \int_{-\infty}^{\infty} dt\, e^{-i(\epsilon_i - \epsilon_f)t} \int d^3 r\, \phi_f^*(\mathbf{r})\, \frac{Z_P}{r_P}\, \phi_i(\mathbf{r}). \tag{3.25}$$

Here, the initial state ϕ_i (for a single-electron problem) is given by a bound-state Coulomb wavefunction. For an initial 1s-state, we have

$$\phi_{1s}(\mathbf{r}) = \frac{1}{\sqrt{4\pi}}\, Z_T^{3/2}\, 2e^{-Z_T r}. \tag{3.26}$$

For excitation, the final state ϕ_f is given by another bound-state wavefunction, e.g., (2.44). For ionization, the final state is defined by the closed-form expression (2.48) for the Coulomb continuum wavefunction in parabolic coordinates or by the corresponding partial-wave expansion. In the latter cases, Eq. (3.24) requires that the wavefunctions are normalized on the energy scale [37, 39], that is

$$\int \phi_E^*(\mathbf{r})\phi_{E'}(\mathbf{r})\, d^3 r = \delta(E - E') \tag{3.27}$$

or, because of the normalization and orthogonality of the angular functions, one requires for the radial functions

$$\int_0^{\infty} R_E(r)R_{E'}(r)\, r^2 dr = \delta(E - E'). \tag{3.28}$$

The behavior of the radial integral is essentially determined by the asymptotic wavefunctions [39]. For a continuum state subject to an attractive Coulomb potential with $\nu_T = Z_T/k$, (where $k = p/\hbar$ is the asymptotic momentum of the emitted electron) we demand that the wavefunction in the limit $\nu_T \to 0$ reduces to a plane wave

$$e^{i\mathbf{k}\cdot\mathbf{r}} = \sum_{l=0}^{\infty} i^l\, (2l + 1)\, j_l(kr)\, P_l(\cos\theta)$$

$$= 4\pi \sum_{l=0}^{\infty} \sum_{m=-l}^{l} i^l\, j_l(kr)\, Y_{lm}^*(\hat{\mathbf{k}})\, Y_{lm}(\hat{\mathbf{r}}). \tag{3.29}$$

Here, $j_l(kr)$ is a spherical Bessel function, $P_l(\cos\theta)$ a Legendre polynomial, while θ is the angle between the unit vectors $\hat{\mathbf{k}}$ and $\hat{\mathbf{r}}$ and $Y_{lm}(\hat{\mathbf{r}})$

is a spherical harmonic. Under this condition, we may define a Coulomb continuum function as [51]

$$
\phi_c(\mathbf{r}) = \sum_{l=0}^{\infty} i^l (2l+1) \phi_l(r) P_l(\cos\theta)
$$

$$
\phi_l(r) = 2^l e^{\frac{1}{2}\pi\nu_{\mathrm{T}}} \frac{\Gamma(l+1-i\nu_{\mathrm{T}})}{(2l+1)!} (kr)^l e^{-ikr}
$$

$$
\times {}_1F_1(l+1+i\nu_{\mathrm{T}}, 2l+2, 2ikr), \tag{3.30}
$$

where the partial waves are expressed by the confluent hypergeometric function ${}_1F_1(a,b,c)$. Normalization on the energy scale yields

$$
\phi_f(r) = (2k/\pi)^{1/2} \phi_c(r). \tag{3.31}
$$

If we wish to adopt the plane-wave Born approximation for ionization, we have to replace $\phi_l(r) \to j_l(kr)$ according to Eq. (3.29).

Although we usually use the straight-line approximation for calculating matrix elements and impact-parameter dependent transition amplitudes $a(b)$, it is meaningful and possible to translate the impact-parameter dependence into a dependence on the deflection angle θ, see, e.g., [2, 52, 53]. Owing to the equivalence of the impact-parameter treatment and the wave picture, see, e.g., [1], it requires essentially a two-dimensional Fourier transform to calculate the scattering amplitude $f(\theta)$ from the transition amplitude $a(b)$ in the center-of-mass system. The result is

$$
\frac{d\sigma}{d\Omega} = \left| \mu v \int_0^{\infty} b^{2i\nu_{\mathrm{PT}}} a(b)\, J_0\left(2\mu v b \sin(\tfrac{1}{2}\theta)\right) b\, db \right|^2, \tag{3.32}
$$

where J_0 is the zero-order Bessel function,

$$
\nu_{\mathrm{PT}} = Z_{\mathrm{P}} Z_{\mathrm{T}}/v \quad \text{and} \quad \mu = M_{\mathrm{P}} M_{\mathrm{T}}/(M_{\mathrm{P}} + M_{\mathrm{T}}). \tag{3.33}
$$

An approximate expression is obtained [2] by disregarding interferences, so that the classical elastic differential b-dependent cross section (1.14) with (1.13) is simply multiplied with the b-dependent transition probability

$$
\frac{d\sigma}{d\Omega} \approx \left(\frac{d\sigma}{d\Omega}\right)_{\mathrm{classical}} |a(b)|^2. \tag{3.34}
$$

The evaluation of the transition amplitude (3.25) is not a simple matter because of the time-dependence of the coordinate $\mathbf{r}_{\mathrm{p}}$. Explicitly, the transition amplitude is given by the expression

$$
a_{fi} = i \int_{-\infty}^{\infty} dt\, e^{-i(\epsilon_i - \epsilon_f)t} \int d^3r\, \phi_f^*(\mathbf{r}) \frac{Z_{\mathrm{P}}}{|\mathbf{r} - \mathbf{R}(t)|} \phi_i(\mathbf{r}). \tag{3.35}
$$

The four-dimensional integration can be partly simplified by decomposing the denominator as

$$\frac{1}{|\mathbf{r} - \mathbf{R}(t)|} = \sum_{L=0}^{\infty} \sum_{M=-L}^{L} \frac{4\pi}{2L+1} \frac{r_<^L}{r_>^{L+1}} Y_{LM}^*(\hat{\mathbf{r}}) \, Y_{LM}(\hat{\mathbf{R}}) \qquad (3.36)$$

with $r_< = \min(r, R)$ and $r_> = \max(r, R)$. Since $\mathbf{r}$ is independent of time, this allows for a partial separation (in each term) of the space integration in Eq. (3.35) referring to the atom and the time integration referring to the trajectory.

3.5   Ionization: The CDW-EIS approximation

In Sec. 3.4, we have outlined the general perturbative treatment of excitation or ionisation of an atom by a charged ion. In this description, the projectile enters merely as a short transient perturbation that does not distort the electron wavefunctions constructed as eigenstates of the target Hamiltonian (2.13). This approach is valid for sufficiently high collision energies.

In the intermediate energy region, it is expected that the distortion of the target wavefunction by the projectile plays a role. Among the various models taking into account the distortion by two Coulomb centers, the apparently most successful one is the continuum-distorted-wave-eikonal-initial-state (CDW-EIS) approximation introduced by Crothers and McCann [54], reviewed in [55] and generalized by Gulyás et al. [56]. This model treats distortions in initial and final channels in an unsymmetric and by no means uniquely determined approximate fashion. It accounts for the long range of the Coulomb potential in both channels. An advantage is that the transition amplitude can be obtained in a closed analytical form for hydrogenic bound and continuum functions. It turns out that an analytical expression can still be obtained for Slater-type orbitals (STO), see Eq. (2.49). Using suitable expansion coefficients [57] within an STO basis, one may represent many-electron targets by Hartree-Fock wavefunctions.

Starting from the electronic Hamiltonian (2.1), a generalized version of the CDW-EIS approximation can be defined by a choice of the initial and final distorted states. The initial distorted wavefunction is

$$\chi_i^+ = \phi_i(\mathbf{r}_T) \, \alpha_i^+(\mathbf{r}_P) \, e^{-i\epsilon_i t} \qquad (3.37)$$

with the target eigenenergy ϵ_i and eigenfunction

$$\phi_i(\mathbf{r}_T) = \frac{1}{r_T} u_{n_i l_i}(r_T) \, Y_{l_i m_i}(\hat{\mathbf{r}}_T), \qquad (3.38)$$

and with the "eikonal" distortion by the projectile defined by

$$\alpha_i^+(\mathbf{r}_P) = e^{-i\nu_P \ln(vr_P + \mathbf{v} \cdot \mathbf{r}_P)}. \tag{3.39}$$

Here, $\nu_P = Z_P/v$ is the Sommerfeld parameter.

The final distorted wavefunction is

$$\chi_f^- = \phi_\epsilon^-(\mathbf{r}_T) \, \alpha_f^-(\mathbf{r}_P) \, e^{-i\epsilon t}, \tag{3.40}$$

where $\phi_\epsilon^-(\mathbf{r}_T)$ is a continuum eigenstate of the target Hamiltonian (2.14) with asymptotic energy $\epsilon = k^2/2$ in the target frame, and the distorting function is

$$\alpha_f^-(\mathbf{r}_P) = N^*(\nu_P) \, {}_1F_1\left[-i\nu_P\,;1\,; -i(pr_P + \mathbf{p} \cdot \mathbf{r}_P)\right], \tag{3.41}$$

with $\mathbf{p} = \mathbf{k} - \mathbf{v}$ being the electron momentum in the projectile frame and $N(\nu_P) = \exp(\pi\nu_P/2)\,\Gamma(1 + i\nu_P)$ a normalization. Note that the factor (3.41) distorting the target continuum function $\phi_\epsilon^-(\mathbf{r}_T)$ in Eq. (3.40) by the projectile charge is the same that distorts the plane wave $\exp(i\mathbf{k} \cdot \mathbf{r})$ in the Coulomb wavefunction (2.48). This is clearly an *ad hoc* ansatz to describe the simultaneous distortion by the target and by the projectile nucleus.

Because the distorted wavefunctions χ_i^+ and χ_f^- cannot be described as solutions of a model Hamiltonian H_{approx}, the resulting transition operator in place of $(H - H_{\text{approx}})$ in Eq. (3.22) can no longer be written as an effective potential, but instead contains derivative operators. Therefore, Eq. (3.22) cannot be applied, and one has to go back to Eq. (3.20) with the substitution (3.21).

The differential cross section is obtained in analogy to Eq. (3.31), see, e.g., [56], by deriving the transverse momentum transfer caused by the projectile from a Fourier transform in the impact-parameter plane and by setting up the momentum balance for the electron.

Doubly differential cross sections

The CDW-EIS model may be applied to the angular and energy distribution of electrons ejected from atoms by bare ions [56]. For multielectron targets, the contributions of the various shells have to be added up. The experimental results do not usually distinguish between single and multiple ionization or other higher-order processes. However, at high collision energies, the single ionization is dominant, so that the present theory is applicable.

Figures 3.1 and 3.2 show calculated and measured doubly differential cross sections (DDCS) as a function of the electron energy. In Fig. 3.1, the target is a multielectron system in which the contributions of the filled shells

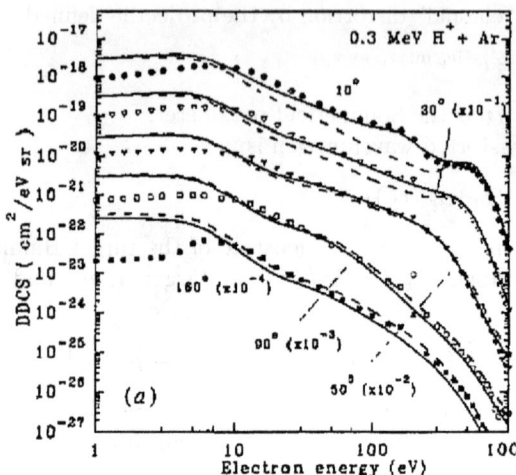

Figure 3.1: Doubly differential cross sections for ionization of Ar by 300 keV proton impact at different observation angles as a function of electron energy. Solid line: CDW-EIS theory; dashed line: first-order Born approximation; symbols: experimental data [58]. From [56].

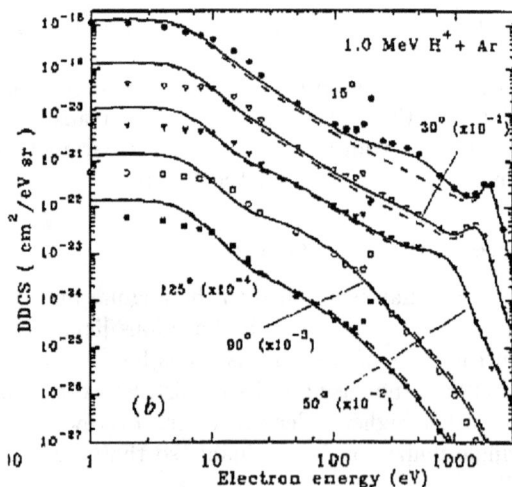

Figure 3.2: Same as Fig. 3.1 but for 1.0 MeV proton impact.

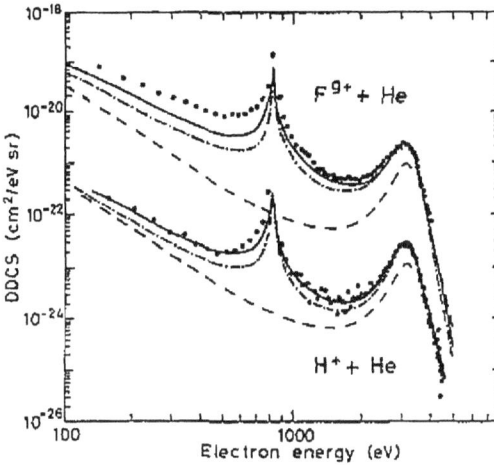

Figure 3.3: Doubly differential cross sections for ionization of He by 1.5 MeV/u protons and F^{9+} impact at observation angle 0° as a function of electron energy. Solid line: CDW-EIS theory [56]; dash-dot line: CDW-EIS [55]; dashed line: first-order Born approximation; symbols: experimental data [59]. From [56].

are summed up. In the high-energy range, we notice a peak or a shoulder, which is denoted as the binary-encounter peak. It arises from a classical binary collision of the projectile with the electron. While for a free electron, we would expect a sharp peak, the momentum distribution of the electron within the target atom gives rise to a finite width, superimposed upon the quantal atomic cross section. For high electron energies, one finds good agreement with the experimental data even for the Born aproximation. For low energies, neither the CDW-EIS nor the Born approximation can explain the experimental results.

In Fig. 3.3, we show the electron spectra at forward angles in collisions of Ar^{9+} and H^+ ions with He. Again we notice the binary encounter peak at high energies, 3000 eV, while at about 800 eV a remarkable cusp structure appears. The cusp structure comes about, if the emitted electron travels with (almost) the same speed and in the same direction as the projectile. This process is sometimes denoted as "charge transfer into the continuum", because the electron is almost captured by the projectile, clearly a two-center effect. Correspondingly, it is rather well described by the CDW-EIS theory (in two versions), which is designed to take into account the distortion effects of the projectile. On the other hand, the first-order Born approximation, which ignores two-center effects, completely fails to describe the cusp structure.

In summary, we see that the CDW-EIS theory gives a rather successful perturbative description of ionization at high collision energies.

In addition to perturbative methods, coupled-channel calculations, see Sec. 2.1, for ionization in energetic $H^+ + H$ collisions have been performed by using a GTO expansion (see (2.60) in Sec. 2.10). Such an approach [60] is unbiased, in principle rigorous, and finds its limitation only in the numerical effort needed for high collison energies. For $H^+ + H$ collisions, satisfactory convergence has been achieved up to 200 keV [60].

Chapter 4

High-energy collisions: Charge transfer

The case of electron capture is more difficult to treat than excitation or ionization. While the theory of direct reactions is essentially straightforward aside from technical details, most of the attention within the field of nonrelativistic ion-atom collisions has been focussed on charge transfer. This subject hence deserves a broader treatment. Indeed, there exists a vast amount of literature on electron capture, for reviews see, e.g., [1, 2, 52, 49]. The reason for the difficulties encountered in the theoretical description, lies in the fact that charge transfer is a rearrangement collision governed by two different atomic Hamiltonians, one for the target and one for the projectile. In the current Chapter, we treat nonradiative capture (NRC). The case of electron capture with the simultaneous emission of a photon, that is, radiative electron capture (REC), is deferred to Chapter 10, which covers both the nonrelativistic and the relativistic situation.

The oldest quantal approach for treating charge transfer is the Oppenheimer-Brinkman-Kramers (OBK) approximation, yielding simple analytical results, which, however, deviate from experimental cross section data by a considerable factor (Sec. 4.1). In order to remedy this defect, higher-order effects have been considered (Sec. 4.2). However, in Sec. 4.3, it is shown that the true cause for the failure lies in the neglect of proper Coulomb boundary conditions taking into account the long range of the Coulomb interaction. If these boundary conditions are imposed, already the lowest-order boundary-corrected formulation, Sec. 4.4, yields realistic capture cross sections. Subsequently, in Sec. 4.5, we describe the continuum-distorted wave (CDW) approximation, which accounts for distortions by the target *and* by the projectile at finite separations. Next,

in Sec. 4.6, we present the eikonal approximation, which leads to realistic
results in a simple analytical form by approximating an important part
of the Coulomb distortion in one channel. As a nonperturbative approach,
the applicability of the coupled-channel method, so successful in low-energy
collisions, see Sec. 2.9, is analyzed in Sec. 4.7 for the high-energy range.
Finally, in Sec. 4.8, we discuss the Thomas double-scattering mechanism.
This effect has found a great deal of attention, because, unexpectedly, the
second-order Born approximation dominates the first.

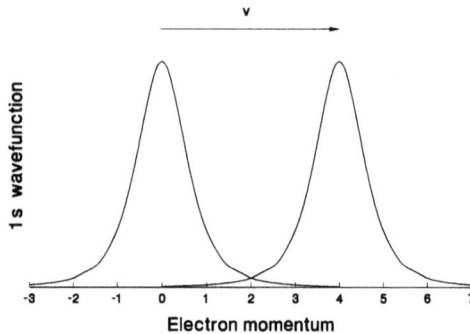

Figure 4.1: Overlap of the 1s momentum wavefunction of a target electron with
the 1s projectile momentum wavefunction shifted by the projectile velocity v.
The overlap of the two curves is essentially responsible for the transfer.

    The essence of the charge transfer process is illustrated in Fig. 4.1. If
an electron "jumps" from the target atom to a fast moving projectile, the
initial wavefunction must have components with high momenta that can be
accommodated in the high-momentum tails of the moving projectile wavefunction in the final state. Figure 4.1 shows the momentum wave function
of a target 1s electron resulting from Eq. (1.21) and the corresponding
projectile wavefunction displaced by the projectile velocity v. The electron transfer will essentially result from the overlap of the two momentum
wavefunctions, however modified by the Coulomb interaction. Already this
simple qualitative picture suggests that the capture cross section will dras-
tically decrease with increasing projectile velocity.

4.1 The Oppenheimer-Brinkman-Kramers approximation

For a quantitative treatment, let us start by writing down the time-dependent Schrödinger equation

$$i\frac{\partial}{\partial t}\Psi(t) = \left(-\tfrac{1}{2}\nabla^2 - \frac{Z_T}{r_T} - \frac{Z_P}{r_P} + \frac{Z_T Z_P}{R}\right)\Psi(t). \tag{4.1}$$

For electron transfer, the initial state is located at the target and the final state at the projectile, so that there is a symmetry between both centers, and – within perturbation theory – one may ask whether Z_P or Z_T gives rise to the perturbation. It is also noted that the internuclear interaction $Z_T Z_P / R$ is included in Eq. (4.1). Indeed, this term might be expected to give contributions to the transition matrix element, because initial and final states are not orthogonal. However, since in the impact-parameter picture the nuclear motion is described by predetermined classical straight-line trajectories, it cannot make a difference for the transition amplitude whether or not the internuclear interaction is present. On the formal side, it is possible to remove the internuclear interaction by a phase transformation of the wavefunction, in analogy to the removal of the asymptotic R-dependencies from Eq. (4.14). It is therefore justified to discard the term $Z_T Z_P / R$ in the following and to retain a purely electronic Hamiltonian.

The earliest quantum treatment of charge transfer has been given in 1928 by Oppenheimer [61] and by Brinkman and Kramers (OBK) [62]. Originally formulated in the wave picture with plane waves for the target and projectile motion, see, e.g. [1], the *equivalent* version within the impact-parameter picture, in analogy to Eq. (3.25), is represented by the amplitude

$$a_{\rm fi} = i \int_{-\infty}^{\infty} dt\, e^{-i(\epsilon_i - \epsilon_f)t} \int d^3r \; \phi_f^*(\mathbf{r_P}) e^{-i\mathbf{v}\cdot\mathbf{r_T} + \frac{i}{2}v^2 t} \frac{Z_P}{r_P}\phi_i(\mathbf{r_T}). \tag{4.2}$$

Note that in contrast to Eq. (3.25), the translation factor (2.19) is included in the moving final-state wavefunction.

It is instructive to give a few steps in the evaluation of Eq. (4.2) because (a) of its simplicity, (b) the same route of evaluation is used for all transfer theories, (c) the main parametric dependencies become exposed, and (d) the crucial relevance of the translation factor is demonstrated.

Adopting $\mathbf{r_P}(t) = \mathbf{r_T} - \mathbf{R}(t)$ with $\mathbf{R}(t) = \mathbf{b} + \mathbf{v}t$, specializing to hydrogen-like 1s wavefunctions, and introducing the Fourier transforms $\tilde{\phi}$

and $\tilde{g}$ by

$$\phi_i(\mathbf{r}_T) = \int \tilde{\phi}_i(\mathbf{q})\, e^{i\mathbf{q}\cdot\mathbf{r}_T}\, d^3q$$

$$\frac{Z_P}{r_P}\, \phi_f(\mathbf{r}_P) = \int \tilde{g}_f(\mathbf{p})\, e^{i\mathbf{p}\cdot\mathbf{r}_P}\, d^3p, \tag{4.3}$$

and inserting into Eq. (4.2), one initially obtains a 10-dimensional integral, however, the space and time coordinates are all moved to the exponent. It follows that the four space and time integrations yield a 4-dimensional δ-function, so that only two of the six momentum integrations are left. Specifically, from the space integration, we obtain $\delta^3(\mathbf{q} - \mathbf{p} + \mathbf{v})$ and from the time integration $\delta(\epsilon_i - \epsilon_f - v^2/2 - p_z v)$. In both cases, terms containing the velocity v arise from the translation factor and are seen to be crucial. The longitudial momentum components are fixed by the δ-functions to

$$p_z = q_- = \frac{\epsilon_i - \epsilon_f}{v} - \frac{v}{2}$$

$$q_z = q_+ = \frac{\epsilon_i - \epsilon_f}{v} + \frac{v}{2} \tag{4.4}$$

with the relation

$$q_-^2 + Z_P^2 = q_+^2 + Z_T^2 \tag{4.5}$$

so that, with the transverse momentum component $\mathbf{q}_\perp$,

$$a_{fi} \propto \frac{1}{v} \int \tilde{g}_f^*(\mathbf{q}_\perp, q_-)\, \tilde{\phi}_i(\mathbf{q}_\perp, q_+)\, e^{i\mathbf{q}_\perp \cdot \mathbf{b}}\, d^2q_\perp . \tag{4.6}$$

According to Eq. (1.21), the momentum wavefunction is

$$\tilde{\phi}_i(\mathbf{q}_\perp, q_+) \propto \frac{Z_T^{5/2}}{(q_\perp^2 + q_+^2 + Z_T^2)^2} . \tag{4.7}$$

Eliminating the potential term Z_P/r_P in $\tilde{g}(\mathbf{q}_\perp, q_-)$ with the aid of the projectile Schrödinger equation, one power in the denominator of $\tilde{g}_f(\mathbf{q}_\perp, q_-)$ is cancelled, so that one has

$$\tilde{g}_f(\mathbf{q}_\perp, q_-) \propto \frac{Z_P^{5/2}}{q_\perp^2 + q_-^2 + Z_P^2} . \tag{4.8}$$

From the relation (4.5), one obtains for the product

$$\tilde{g}_f^*(\mathbf{q}_\perp, q_-)\, \tilde{\phi}_i(\mathbf{q}_\perp, q_+) \propto \frac{Z_P^{5/2} Z_T^{5/2}}{(q_\perp^2 + q_+^2 + Z_T^2)^3} . \tag{4.9}$$

After taking the modulus square of (4.6) and integrating over the impact parameter plane, one obtains another δ-function $\delta^2(\mathbf{q}_\perp - \mathbf{q}'_\perp)$, so that one has a single integration over the transverse momenta. Supplementing the constants and replacing the final 1s shell with a complete principal n-shell, one finally derives the well-known formula, see e.g. [49],

$$\sigma^{\mathrm{OBK}}(1s-n) = \frac{2^8 \pi Z_P^5 Z_T^5}{5n^3 v^2 (Z_T^2 + q_+^2)^5} \tag{4.10}$$

with the asymptotic high-energy behavior

$$\sigma^{\mathrm{OBK}}(1s-n) \longrightarrow \frac{2^{18} Z_P^5 Z_T^5}{5n^3 v^{12}} \qquad \text{for} \qquad v \to \infty. \tag{4.11}$$

The sharp decrease of the cross section with velocity, see also Sec. 4.8, is caused by the translation factor while the asymptotic Z-dependence arises from the Fourier transforms of the wavefunctions. The basic asymptotic charge- and energy- dependence expressed by Eq. (4.11) is common to all (nonrelativistic) capture theories, except, strangely, for second-order theories. The dependence with n^{-3} in Eq. (4.11) is often used to estimate the effects of higher shells, whenever in a theoretical description they are not included explicitly. Adopting the n^{-3}-rule, it is reasonable for a rough estimate to multiply the cross section for capture into the 1s-shell by a factor of $\sum_{n=1}^{\infty} n^{-3} = 1.202$ in order to account for the contributions from all higher shells.

4.2 The Jackson-Schiff and the Strong-Potential Born approximations

Oppenheimer as well as Brinkman and Kramers argued on physical grounds that the term $Z_P Z_T / R$ as a perturbation in Eq. (4.1) should give no contribution. However, because of nonorthogonality of $\phi_f(\mathbf{r}_P)$ and $\phi_i(\mathbf{r}_T)$, the term $Z_P Z_T / R$ or any arbitrary function $W(R)$ will contribute in (4.2). Later, Jackson and Schiff (JS) [63] included this term. It turns out that both "Born approximations", OBK and JS, give grossly incorrect results (except for H$^+$ on H, see Sec. 4.4), in striking contrast to results for excitation and ionization. The failure is particularly striking for large Z_P, which became experimentally accessible at sufficiently high energies only in the 1970ies and 1980ies. To make things worse, it turned out that the inclusion of the second-order term (OBK2) in the OBK expansion (which actually would be justified theoretically only for short-range interactions, that is, potentials decaying faster than $1/r$) led to a deterioration of the

agreement with experimental total cross sections in the intermediate energy range. Only in the high-energy regime, the OBK2 yields qualitatively
correct results, see Sec. 4.8.

In this situation, it was suggestive to study even higher-order terms in
the OBK perturbation series. This line of approach was implemented by
Macek and co-workers [64, 65, 66, 67] in a model known as "Strong Potential Born" (SPB) approximation. The SPB approximation is derived from
an earlier approach developed by Briggs [68] and denoted as "Plane-Wave
Impulse Approximation". It can be regarded as a multiple-scattering vari-
ant of the OBK theory in so far as one of the two Coulomb potentials,
preferably the weaker one (that is, with the smaller charge number), is
treated to first order and the other one to all orders of the OBK perturbation theory, albeit in an approximate fashion. In this way, the model
became a good approximation for asymmetric ($Z_P \neq Z_T$) collisions but
not for the prototype $H^+ + H$ and $H^+ + He$ reactions.

The SPB approximation evolved, when certain important terms omitted in the plane-wave impulse approximation were supplemented in the
systematic expansion in powers of the weaker potential. Because the numerical evaluation of the second-order SPB amplitude is a difficult task,
additional simplifying (mathematical) approximations, the so-called *peaking approximations* were introduced in numerical applications. The *peaked*
SPB approximation [65] was found to yield reasonable agreement with total electron transfer cross sections for various asymmetric collisions. As
a consequence, in the 1980ies, the SPB approximation has attracted wide
interest and attention.

However, it is known that the peaking approximation often introduces
large errors [3]. This was born out in 1985, when Dewangan and Eichler
[69] demonstrated that the *exact* SPB amplitude embodies a singularity,
which disappears only if the peaking approximation is applied. This difficulty is not strictly confined to the SPB approximation but also emerges in
the standard T-matrix element of the OBK formulation [70]. The reason
has been traced to the long range of the Coulomb interaction as has been
first pointed out in [69, 71] and was summarized in [49]. In fact, the perturbation expansion (3.7) is invalid for long-range potentials. Indeed, even
at asymptotic separations, the Coulomb continuum wavefunction does not
approach a plane wave.

4.3 Coulomb boundary conditions and gauge transformations

It is instructive to analyze the effects of the long range of the Coulomb interaction, because they constitute a general feature of ion-atom collisions. In Fig. 4.2, we illustrate the situation for the outgoing and the incoming

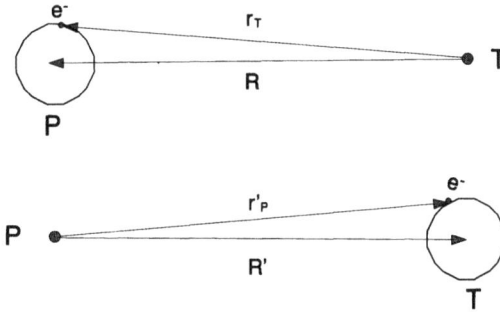

Figure 4.2: The projectile atom seen from a distant target nucleus ($t \to \infty$), and the target atom seen from a distant projectile nucleus($t \to -\infty$). For non-relativistic collisions, the primes should be disregarded.

channels, when the collision partners are far from each other. Starting explicitly from the Schrödinger equation (4.1), we have

$$
\begin{array}{lll}
\text{for} & t \to -\infty & r_{\mathrm{P}} \to R \\
\text{for} & t \to \infty & r_{\mathrm{T}} \to R.
\end{array}
\tag{4.12}
$$

Hence the exact solutions $\Psi_i^+(t)$ and $\Psi_f^-(t)$ satisfy the boundary conditions

$$
\begin{aligned}
\lim_{t \to -\infty} \Psi_i^+(t) &= \Phi_i^\infty(t) \\
\lim_{t \to \infty} \Psi_f^-(t) &= \Phi_f^\infty(t)
\end{aligned}
\tag{4.13}
$$

with

$$
\begin{aligned}
i\frac{\partial}{\partial t}\Phi_i^\infty(t) &= \left(-\tfrac{1}{2}\nabla^2 - \frac{Z_{\mathrm{T}}}{r_{\mathrm{T}}} - \frac{Z_{\mathrm{P}}}{R}\right)\Phi_i^\infty(t) \\
i\frac{\partial}{\partial t}\Phi_f^\infty(t) &= \left(-\tfrac{1}{2}\nabla^2 - \frac{Z_{\mathrm{T}}}{R} - \frac{Z_{\mathrm{P}}}{r_{\mathrm{P}}}\right)\Phi_f^\infty(t).
\end{aligned}
\tag{4.14}
$$

Denoting the Sommerfeld parameters as

$$\nu_{\mathrm{P}} = Z_{\mathrm{P}}/v \qquad \text{and} \qquad \nu_{\mathrm{T}} = Z_{\mathrm{T}}/v, \tag{4.15}$$

we may remove the asymptotic terms proportional to $1/R$ from Eq. (4.14) by a phase transformation[†], so that

$$\begin{aligned}
\Phi_{\mathrm{i}}^{\infty}(t) &= \Phi_{\mathrm{i}}(\mathbf{r}_{\mathrm{T}}, t) \, e^{-i\nu_{\mathrm{P}} \ln(vR - v^2 t)} && \text{for} && t \to -\infty \\
\Phi_{\mathrm{f}}^{\infty}(t) &= \Phi_{\mathrm{f}}(\mathbf{r}_{\mathrm{P}}, t) \, e^{i\nu_{\mathrm{T}} \ln(vR + v^2 t)} && \text{for} && t \to \infty
\end{aligned} \tag{4.16}$$

where, in analogy to Eq. (2.17), the time-dependent atomic bound-state wavefunctions are written as

$$\begin{aligned}
\Phi_{\mathrm{i}}(\mathbf{r}_{\mathrm{T}}, t) &= \phi_{\mathrm{i}}(\mathbf{r}_{\mathrm{T}}) \, e^{-i\epsilon_i t} \\
\Phi_{\mathrm{f}}(\mathbf{r}_{\mathrm{P}}, t) &= \phi_{\mathrm{f}}(\mathbf{r}_{\mathrm{P}}) \, e^{i\mathbf{v}\cdot\mathbf{r}_{\mathrm{T}} - \frac{i}{2} v^2 t} \, e^{-i\epsilon_f t},
\end{aligned} \tag{4.17}$$

with ϕ_{i} and ϕ_{f} being the unperturbed stationary solutions of the atomic Schrödinger equations in their own coordinate systems. For the moving projectile, the electron translation factor (2.19) is included. The phase transformation (4.16), which takes into account the Coulomb boundary conditions, may also be regarded as a *gauge transformation*.

In order to study the effect of this transformation on the transition amplitude, we start from the *exact* definitions as projections of the asymptotic initial or final states on the exact outgoing- or incoming- wave solutions, respectively, in analogy to Eqs. (2.21) and (3.16), see [2, 49, 3]. However, for rearrangement collisions, we have to treat the *post* and *prior* forms separately, because, in contrast to excitation and ionization, the Hamiltonians defining initial and final states are different. We hence have two versions of the transition amplitude

$$\begin{aligned}
a_{\mathrm{fi}}^{(+)} &= \lim_{t \to \infty} \langle \Phi_{\mathrm{f}}^{\infty} | \Psi_{\mathrm{i}}^{+} \rangle && \text{post form} \\
a_{\mathrm{fi}}^{(-)} &= \lim_{t \to -\infty} \langle \Psi_{\mathrm{f}}^{-} | \Phi_{\mathrm{i}}^{\infty} \rangle && \text{prior form.}
\end{aligned} \tag{4.18}$$

Since for electron capture, the overlap $\langle \Phi_{\mathrm{f}}^{\infty} | \Psi_{\mathrm{i}}^{+} \rangle$ vanishes for $t \to -\infty$, the *post form* can be written as

$$\begin{aligned}
a_{\mathrm{fi}}^{(+)} &= -i \int_{-\infty}^{\infty} dt \left(\left\langle \Phi_{\mathrm{f}}^{\infty} \left| i\frac{\partial}{\partial t} \Psi_{\mathrm{i}}^{+} \right\rangle - \left\langle i\frac{\partial}{\partial t} \Phi_{\mathrm{f}}^{\infty} \right| \Psi_{\mathrm{i}}^{+} \right\rangle \right) \\
&= -i \int_{-\infty}^{\infty} dt \left\langle \left(H - i\frac{\partial}{\partial t} \right) \Phi_{\mathrm{f}}^{\infty} \middle| \Psi_{\mathrm{i}}^{+} \right\rangle.
\end{aligned} \tag{4.19}$$

[†]Various existing forms of the phase are equivalent, if they differ only by an additive constant phase. Note, for example, that $b^2 = (R + vt)(R - vt)$ and hence $\ln(R + vt) = -\ln(R - vt) + \text{const}.$

Correspondingly, we obtain the *prior form* as

$$a_{\mathrm{fi}}^{(-)} = -\mathrm{i} \int_{-\infty}^{\infty} \mathrm{d}t \left\langle \Psi_{\mathrm{f}}^{-} \left| H - \mathrm{i} \frac{\partial}{\partial t} \right| \Phi_{\mathrm{i}}^{\infty} \right\rangle. \tag{4.20}$$

The equations (4.19) and (4.20) are both exact and hence identical, provided that Ψ_{i}^{+} and Ψ_{f}^{-} are exact solutions. They constitute the basis for all distorted wave Born approximations (DWBA) for electron transfer.

If approximate wavefunctions are substituted for the exact ones, the identity does not hold in all cases. This is denoted as *post-prior discrepancy*. Similarly as in Sec. 3.3 for direct reactions, the physical meaning of the equations is the following: In the post form, the operator $(H - \mathrm{i}\partial/\partial t)$ acting on $\Phi_{\mathrm{f}}^{\infty}$ yields the potential term, i.e., the *residual interaction* that is *not* included in $\Phi_{\mathrm{f}}^{\infty}$ at $t \to \infty$. For charge transfer, this is the electron-target interaction. A corresponding argument holds for the prior form.

The Born approximation is obtained by substituting unperturbed initial and final states for the exact ones. A variety of distorted-wave Born approximations can be constructed by inserting distorted states, in which a suitable part of the Hamiltonian is included in $\Phi_{\mathrm{i}}^{\infty}$ and $\Phi_{\mathrm{f}}^{\infty}$. The operator $(H - \mathrm{i}\partial/\partial t)$ then produces the perturbation left over from the assumed distortion.

With the aim of taking into account the Coulomb boundary conditions, we next insert $\Phi_{\mathrm{f}}^{\infty}$ from Eq. (4.16) into the post form of the transition amplitude. Equation. (4.19) then yields the amplitude

$$a_{\mathrm{fi}}^{(+)} = \mathrm{i} \int_{-\infty}^{\infty} \mathrm{d}t \left\langle \Phi_{\mathrm{f}}(\mathbf{r}_{\mathrm{P}}, t)\, \mathrm{e}^{\mathrm{i}\nu_{\mathrm{T}} \ln(vR + v^2 t)} \left| \frac{Z_{\mathrm{T}}}{r_{\mathrm{T}}} - \frac{Z_{\mathrm{T}}}{R} \right| \Psi_{\mathrm{i}}^{+} \right\rangle, \tag{4.21}$$

which is still exact. We note that by taking into account the long-range Coulomb interaction in the phase factor, the residual interaction in Eq. (4.21) for $t \to \infty$ is of *short range*. In general, the phase transformations (4.16) lead to the replacements

$$\frac{Z_{\mathrm{T}}}{r_{\mathrm{T}}} \longrightarrow \frac{Z_{\mathrm{T}}}{r_{\mathrm{T}}} - \frac{Z_{\mathrm{T}}}{R}$$

$$\frac{Z_{\mathrm{P}}}{r_{\mathrm{P}}} \longrightarrow \frac{Z_{\mathrm{P}}}{r_{\mathrm{P}}} - \frac{Z_{\mathrm{P}}}{R}. \tag{4.22}$$

The short range implies that standard perturbation theory is applicable. The singularities of the asymptotic terms Z_{T}/R and Z_{P}/R at $R = 0$, that is at $b = 0$ and $t = 0$, are innocuous in the matrix elements, and it is possible to modify the asymptotic terms at finite values of R without losing the benefit of producing a short-range interaction similarly as in Eq. (4.22).

4.4   The boundary-corrected first Born (B1B) approximation

We are now in a position to build a first-order approximation on the exact equation (4.21). If we replace the exact wavefunction $\Psi_i^+ \to \Phi_i^\infty$, we obtain a first-order expression denoted as the "boundary-corrected first Born approximation" (B1B) introduced by Dewangan and Eichler[†] [71, 72], in the post form

$$a_{\mathrm{fi}}^{\mathrm{B1B}+} = \mathrm{i} \int_{-\infty}^{\infty} \mathrm{d}t \left\langle \Phi_{\mathrm{f}}\, \mathrm{e}^{\mathrm{i}\nu_{\mathrm{T}}\ln(vR+v^2 t)} \left| \frac{Z_{\mathrm{T}}}{r_{\mathrm{T}}} - \frac{Z_{\mathrm{T}}}{R} \right| \Phi_{\mathrm{i}}\, \mathrm{e}^{-\mathrm{i}\nu_{\mathrm{P}}\ln(vR-v^2 t)} \right\rangle,$$

(4.23)

and correspondingly in the prior form

$$a_{\mathrm{fi}}^{\mathrm{B1B}-} = \mathrm{i} \int_{-\infty}^{\infty} \mathrm{d}t \left\langle \Phi_{\mathrm{f}}\, \mathrm{e}^{\mathrm{i}\nu_{\mathrm{T}}\ln(vR+v^2 t)} \left| \frac{Z_{\mathrm{P}}}{r_{\mathrm{P}}} - \frac{Z_{\mathrm{P}}}{R} \right| \Phi_{\mathrm{i}}\, \mathrm{e}^{-\mathrm{i}\nu_{\mathrm{P}}\ln(vR-v^2 t)} \right\rangle.$$

(4.24)

Returning to Eq. (4.1), it is seen that, in analogy to Eq. (4.16), the inclusion of the internuclear potential $Z_{\mathrm{P}} Z_{\mathrm{T}}/R$ would contribute only a phase factor $(bv)^{\mathrm{i}2 Z_{\mathrm{P}} Z_{\mathrm{T}}/v}$, which does not alter the capture amplitude for fixed values of b and v. In the special case of proton-hydrogen collisions ($Z_{\mathrm{P}} = Z_{\mathrm{T}} = 1$), we have $\nu_{\mathrm{P}} = \nu_{\mathrm{T}} = 1/v$, so that the phase factors occurring in Eqs. (4.23) and (4.24) can be combined to give $(bv)^{-\mathrm{i}2/v}$, a quantity independent of the variables of integration, which is cancelled by the phase factor resulting from the internuclear potential. This is the reason, why the Jackson-Schiff approximation discussed in Sec. 4.2 yields realistic results for p + H collisions but fails everywhere else [49].

For the total 1s – 1s B1B cross sections, closed-form expressions in terms of simple algebraic functions have been derived by Dubé et al. [73]. For asymptotically high energies one obtains the result

$$\sigma^{\mathrm{B1B}}(1s-1s) \longrightarrow 0.661\, \sigma^{\mathrm{OBK}}(1s-1s) \qquad \text{for} \qquad v \to \infty, \quad (4.25)$$

where the asymptotic OBK cross section is given by Eq. (4.11) for $n = 1$.

The important point in the preceding discussion is the observation that for a proper description of charge exchange, it is indispensable to take

---

[†]Similar expressions have also been written down by Belkić et al. [52], however in a different context and without much consequence.

the Coulomb boundary conditions, that is, the long-range nature of the Coulomb interaction explicitly into account. The B1B approximation is just the simplest member of a large family of approaches satisfying these requirements. It turns out that one obtains good agreement with experimental data, see Fig. 4.3. Corrections for screening in a multielectron atom [74] and the impact-parameter dependence of the transition amplitude [75] have also been considered. We now wish to summarize a few points.

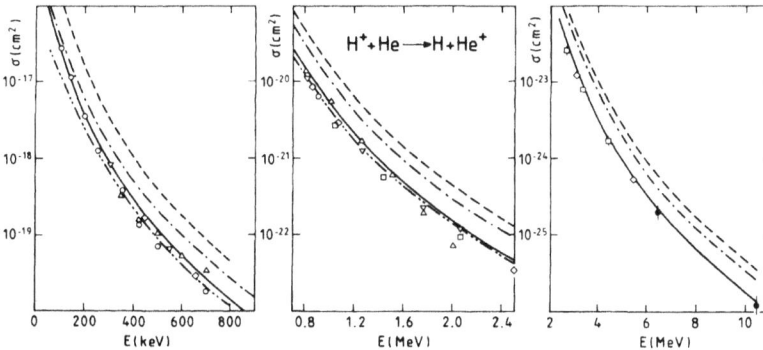

Figure 4.3: Electron-capture cross sections for $H^+ + He \rightarrow H + He^+$. Theoretical results: Solid curve: B1B; dash-dot curve: DWB1B; dashed curve: OBK; dash-dot-dot curve: eikonal, see Sec. 4.6. In the right figure, corresponding to the highest energy, the eikonal curve is indistiguishable from the B1B curve. Points are experimental data. From [75].

- The B1B approximation is the first term of a divergence-free and hence consistent Born expansion.

- The B1B approximation is post-prior symmetric. The same is also true for the OBK approximation, but this feature by itself cannot be regarded as a sufficient criterion for validity.

- In principle, boundary-corrected states should be used also for excitation and ionization. However, since initial and final states are eigenstates of the same Hamiltonian, the phase factors are identical and hence drop out. For the same reason, the $1/R$ terms do not contribute, owing to orthogonality.

- In coupled-channel calculations, see Secs. 2.1 and 2.9, the use of a boundary-corrected basis presumably should improve convergence.

However, since the rapidly oscillating phase factors are difficult to treat numerically, it may be profitable to include more but simpler wavefunctions.

- The amplitudes (4.23) and (4.24) are the simplest versions for a theory taking into account the long range of the Coulomb potential. One may easily construct modifications of the R-dependence at *finite* R, removing the singularity at $R = 0$, thus taking into account screening effects of the electron cloud in an atom [74]. Interestingly, the corresponding modified DWB1B approximation deteriorates the good agreement with experimental data at high energies [75], see Fig. 4.3.

- Second-order calculations (B2B approximation) have also been performed, see [49].

4.5   The continuum-distorted wave (CDW) approximation

The "continuum-distorted wave" (CDW) theory [76] represents a multiple-scattering model in which higher-order perturbation terms are included. This successful approach has received most attention among all capture theories, see also Sec. 3.5 for ionization. It is usually considered as the best-justified approach.

In the CDW approximation, the distortion does not only take into account the asymptotic behavior as in B1B, but also embodies some Coulomb effects at finite separations. Similarly as in B1B, we have

$$\Psi_i^+ = \Phi_i(t)X_i(t) \quad \text{with} \quad X_i(t) \to e^{-i\nu_P \ln(vR - v^2 t)}$$
$$\text{for} \quad t \to -\infty$$
$$\Psi_f^- = \Phi_f(t)X_f(t) \quad \text{with} \quad X_f(t) \to e^{i\nu_T \ln(vR + v^2 t)}$$
$$\text{for} \quad t \to \infty, \tag{4.26}$$

but for finite values of t, with the functions Φ_i, Φ_f given in Eq. (4.17), the modifying factors are

$$X_i(t) = N(\nu_P) \, _1F_1[i\nu_P; 1; i(vr_P + \mathbf{v} \cdot \mathbf{r}_P)]$$

$$X_f(t) = N^*(\nu_T) \, _1F_1[-i\nu_T; 1; -i(vr_T + \mathbf{v} \cdot \mathbf{r}_T)] \tag{4.27}$$

with $N(\nu) = \Gamma(1 - i\nu)\exp(\pi\nu/2)$. Here, at a finite separation, $X_i(t)$ represents the Coulomb distortion of the bound target wavefunction by the projectile, which absorbs an electron with relative velocity $-\mathbf{v}$, see

Eq. (2.47), while $X_f(t)$ describes the Coulomb distortion of the bound projectile wavefunction by the target, which emits an electron with velocity
$\mathbf{v}$, see Eqs. (2.48) and (3.41). The compact description of the distortion is
possible, because the nonrelativistic Schrödinger equation with a Coulomb
potential can be separated in parabolic coordinates.

Adopting the expressions (4.27), also the residual interaction is changed.
We just write down the *post* amplitude

$$
a_{fi}^{\text{CDW}+}(b) = iN(\nu_{\text{T}})N(\nu_{\text{P}}) \int_{-\infty}^{\infty} dt \int d^3 r_{\text{T}} e^{-i\mathbf{v}\cdot\mathbf{r}_{\text{T}} + \frac{i}{2}v^2 t} e^{-i(\epsilon_i - \epsilon_f)t}
$$

$$
\times \phi_i(\mathbf{r}_{\text{T}}) \, {}_1F_1[i\nu_{\text{P}}; 1; i(vr_{\text{P}} + \mathbf{v}\cdot\mathbf{r}_{\text{P}})]
$$

$$
\times \{\nabla_{r_{\text{T}}} \phi_f^*(\mathbf{r}_{\text{P}}) \cdot \nabla_{r_{\text{T}}} \, {}_1F_1[i\nu_{\text{T}}; 1; i(vr_{\text{T}} + \mathbf{v}\cdot\mathbf{r}_{\text{T}})]\}.
$$
(4.28)

The transition operator here has no longer the appearance of a potential
and is very difficult to evaluate. Contributions from all orders in $Z_{\text{P}}/r_{\text{P}}$ and
$Z_{\text{T}}/r_{\text{T}}$ are included. A problem is that CDW functions are not normalized
at finite R and overestimate the cross section at lower velocities. For a
more detailed discussion, see [52, 49].

4.6   The eikonal approximation

Another approach, which gives surprisingly good results for total cross
sections and yet can be evaluated as easily as the (incorrect) OBK approximation, is the eikonal approximation worked out by Chan and Eichler
[77, 78, 79] and by Eichler [80].

The post-form amplitude is given by

$$
a^{\text{EA}+}(b) = i \int_{-\infty}^{\infty} dt \int d^3 r_{\text{T}} \, \Phi_f^*(t) \frac{Z_{\text{T}}}{r_{\text{T}}} \Phi_i(t) \, e^{-i\nu_{\text{P}} \ln(vr_{\text{P}} + \mathbf{v}\cdot\mathbf{r}_{\text{P}})} \qquad (4.29)
$$

and the prior form by

$$
a^{\text{EA}-}(b) = i \int_{-\infty}^{\infty} dt \int d^3 r_{\text{T}} \, e^{-i\nu_{\text{T}} \ln(vr_{\text{T}} + \mathbf{v}\cdot\mathbf{r}_{\text{T}})} \Phi_f^*(t) \frac{Z_{\text{P}}}{r_{\text{P}}} \Phi_i(t). \qquad (4.30)
$$

Both versions are *not* identical. The approach also does not satisfy
Coulomb boundary conditions in both channels, so it has theoretical defects. Note that the arguments of the logarithm in the exponent do not

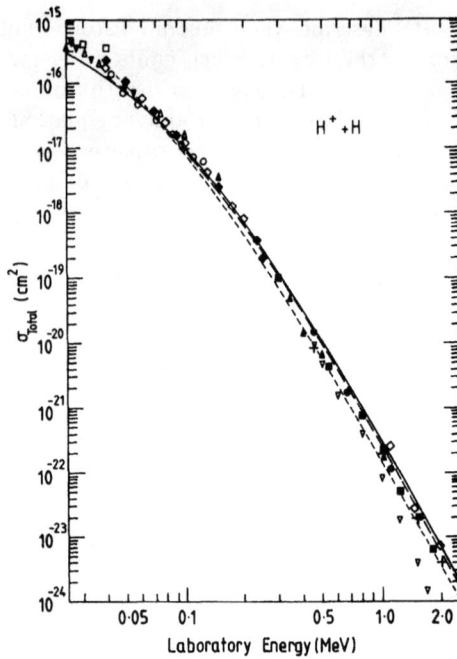

Figure 4.4: Total cross sections for electron capture by protons from atomic hydrogen. Theoretical results: Solid curve: B1B cross section obtained by explicitly summing over the contributions from all subshells up to the L-shell of H and by accounting for capture into the remaining excited states by the n_f^{-3} rule [see Sec. 4.1]; short dash: CDW cross section [52]; long dash: cross section in the eikonal approximation. Points are experimental data. From [49].

depend on time but on the electronic coordinate. Since, e.g., in the prior form, with $\mathbf{r}_P$ kept fixed

$$\exp\big(-i\nu_T \ln(r_T + z_T)\big) \propto \exp\Big(-i \int_t^\infty \frac{Z_T}{r_T} dt'\Big), \tag{4.31}$$

the phase factor is represented by an eikonal integral describing the interaction of the electron with the target nucleus from the time of capture to infinity. The transition amplitudes (4.30) and (4.29) have, therefore, features of a multiple-scattering or nonperturbative approach. In the prior form, Z_T is contained to all orders, in the post form Z_P. The analytical evaluation can be achieved by Fourier transforming into momentum space, similarly as in Sec. 4.1. Using an expression given by Fock [10], one may calculate the capture cross section from a 1s-state into a complete n-shell [77, 78]. As a result, one obtains in the prior form

$$\sigma^{EA-}(1s-n) = \alpha_n(Z_P, Z_T, v)\,\sigma^{OBK}(1s-n), \qquad (4.32)$$

where the OBK cross section σ^{OBK} is given in Eq. (4.10), and the reduction factor is

$$\alpha_n(Z_P, Z_T, v) = \frac{\nu_T\,\pi}{\sinh(\nu_T\pi)}\,e^{-2\nu_T\,\arctan(q_+/Z_T)}$$

$$\times\left[\tfrac{23}{48} + (\tfrac{1}{6}Z_T^2 + \tfrac{5}{6}\epsilon)v^{-2} + \tfrac{5}{12}\,\epsilon^2\,v^{-4}\right] \qquad (4.33)$$

with $\epsilon = \epsilon_f - \epsilon_i$ and $q_+ = v/2 - \epsilon/v$, see Eq. (4.4). The coefficient α_n reduces the OBK cross section by a factor 3 to 4, just what is needed to achieve agreement with experimental total cross sections. The asymptotic behavior at high energies is for $n = 1$

$$\sigma^{EA}(1s-1s) \longrightarrow \tfrac{23}{48}\,\sigma^{OBK}(1s-1s) \qquad \text{for} \qquad v \to \infty, \qquad (4.34)$$

both in the post and in the prior form, with the asymptotic OBK cross section given by Eq. (4.11). Similarly, cross sections into and from subshells, $(1s - n_f, l_f)$ [79] and $(n_i, l_i - n_f, l_f)$ [80], have been evaluated and are given in a rather simple analytical form. Computer codes are available that lend themselves for fast and realistic estimates, even for high n and l.

It should be mentioned that the eikonal approach does not satisfy Coulomb boundary condition and hence is not fully justifiable on theoretical grounds. However, the agreement of results with better justified theories (like the B1B and CDW approximation) and with experiment is surprisingly good. An example is given in Fig. 4.4 for the prototype case of $H^+ + H$ collisions. There exist many more approaches and modifications. For a summary, see [49].

4.7 Coupled-channel calculations for transfer at high energies

There are essentially two possible ways to extend the first-order theory for electron transfer. One is a higher-order perturbation theory, the other is the coupled-channel method. The former takes into account *all couplings* up to some *finite order*, the latter takes into account *all orders* of couplings within a limited space spanned by a *finite set* of basis functions. In this sense, the two approaches are complementary.

As has been demonstrated in Sec. 2.9, the two-center coupled-channel method is one of the most reliable theoretical approaches to electron excitation as well as to capture in slow and intermediate-energy ion-atom collisions. We have to examine whether this method can be carried over to energetic collisions. From Eq. (2.38) we have the expansion

$$\Psi(\mathbf{r}, t) = \sum_k a_k(t)\phi_k(\mathbf{r}_{\mathrm{T}})e^{-i\epsilon_k t} + \sum_{k'} a_{k'}(t)\phi_{k'}(\mathbf{r}_{\mathrm{P}})\,e^{i\mathbf{v}\cdot\mathbf{r}-\frac{1}{2}v^2 t}e^{-i\epsilon_{k'} t},$$

(4.35)

where $k \to nlm$ and $k' \to n'l'm'$ denote the sets of quantum numbers at target and projectile, respectively. One particular merit of this expansion lies in the fact that there is no difference between the transition probabilities obtained with or without Coulomb boundary conditions. This is so, because any function of $R(t)$ depends only on time (not on the coordinates), as for example in Eq. (4.14) or (4.21), therefore can be taken out of the transition matrix element and absorbed into the expansion coefficients $a_k(t)$ and $a_{k'}(t)$ of Eq. (4.35). We hence may say that Coulomb boundary conditions are always satisfied [75].

Unfortunately, the coupled-channel method finds its limitation at higher projectile velocities, when the interaction matrix elements and their integrands assume an increasingly oscillatory behavior, mainly caused by the translation factor in Eq. (4.35). An accurate numerical evaluation then becomes exceedingly difficult. Applications above several hundred keV collision energy are therefore scarce.

In order to extend the applicability of the coupled-channel method to higher energy and to remedy the numerical difficulties in the evaluation of matrix elements, it has been proposed to use the analyticity properties of a Gaussian basis, see Sec. 2.10 [42, 60]. In this approach, the eigenfunctions ϕ_{klm} are expanded as

$$\phi_{nlm}(\mathbf{r}) = \sum_{\nu} C_{\nu}^{(nl)} e^{-\alpha_\nu r^2}\, r^l\, Y_{lm}(\hat{\mathbf{r}}),$$

(4.36)

where the solid harmonics $r^l Y_{lm}(\hat{\mathbf{r}})$ can be written in Cartesian coordinates.

By diagonalizing the atomic Hamiltonian for each center, one obtains bound states as well as a large number of states with positive eigenvalues. The latter are interpreted as pseudostates which, by superposition, approximate the oscillating continuum wave functions up to a maximum range $\bar{r}_{\max} = (\alpha_\nu)^{-1/2}$ and then fall off. It has been found [42] that for high-energy charge transfer in $H^+ + H$ collisions (see Sec 4.8) a very high precision of about 10^{-14} is needed for the energy levels in the diagonalization procedure, because even a small contamination with continuum states, gives rise to spurious contributions to bound-state capture. The reason is that ionization cross sections at these energies are by orders of magnitude larger than capture cross sections. In the two-center matrix elements, the translation factor $\exp(i\mathbf{v} \cdot \mathbf{r})$ enters, see Eqs. (2.40) and (2.41), and an analytical evaluation in terms of elementary functions is no longer possible. It is then necessary to introduce a complex auxiliary function related to the incomplete gamma function, which again has to be evaluated with a very high accuracy [42]. It is seen that for the possibility of analytical evaluation, one has to pay with the requirement for a very high precision, at least in describing high-energy collisions.

Besides reproducing the Thomas peak in the differential cross section, see Sec. 4.8, the following general features are found [81]. Two-center coupled-channel calculations for $H^+ + H(1s)$ collisions, which contain *only bound states* in the expansion (4.35) at target and projectile, rapidly tend to the OBK cross section as the collision energy increases. They are hence unable to describe the experimental data. Only the inclusion of continuum states in the expansion leads to a reduction factor in the cross section compared to the OBK results, so that total cross sections come close to those for the CDW approximation (see Sec. 4.5) and are similar to those for the B1B and eikonal results (see Secs. 4.4 and 4.6). As a consequence, the experimental cross sections can be reproduced.

This analysis shows that the inclusion of continuum states in coupled-channel calculations is crucial for describing bound-to-bound transitions in high-energy charge transfer.

4.8 The Thomas double-scattering mechanism

The high-energy behavior of charge transfer cross sections has found a great deal of attention. Already in 1927, L.H. Thomas proposed a classical mechanism for a bound-to-bound transition by two correlated binary collisions [82], see Fig. 4.5. In the first collision, the projectile nucleus P deflects the target electron with the speed v into an angle close to $60°$ (with respect to the projectile velocity $\mathbf{v}$). In the second collision, the struck electron is scattered elastically by the target nucleus, again with $60°$ in such a way that it emerges with nearly zero speed with respect to P and now can be bound by the projectile's Coulomb potential. Thomas showed that within this mechanism,

$$\sigma_{\text{classical}} \propto v^{-11}. \tag{4.37}$$

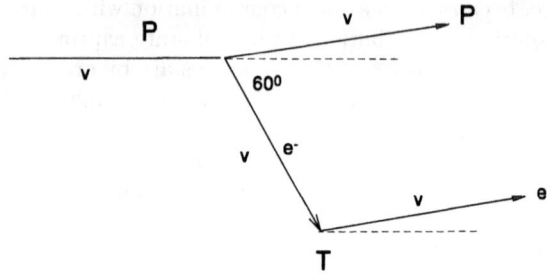

Figure 4.5: Thomas double scattering. The effective mass for $H^+ + H$ collisions is taken as the proton mass M_p.

The projectile recoils by a small angle, the Thomas angle

$$\Theta_{Th} = \frac{m_e}{\mu} \sin 60° = \frac{\sqrt{3}}{2} \frac{m_e}{\mu}, \tag{4.38}$$

in the center-of-mass system, where μ, defined in Eq. (3.33), is the reduced mass of target and projectile [82, 83]. For $H^+ + H$ collisions, $\Theta_{Th} \approx 0.475$ mrad.

On the other hand, the quantum cross section, see Eq. (4.11), behaves asymptotically as $\sigma(1s - 1s) \propto v^{-12}$ and decreases more rapidly for higher values of l. This is true for the OBK, the B1B, the eikonal and all first-order approaches. The discrepancy was fist resolved by Drisko in 1955 [84], who introduced a second-order Born approximation (OBK2) reflecting the classical double collision. For $H^+ + H$ collisions, he used the second-order Jackson-Schiff approximation with plane waves, which for $Z_P = Z_T = 1$ is identical to the second-order boundary-corrected B2B0 approximation (see, e.g., [49]). He found

$$\sigma^{B2B0}(1s - 1s) \approx \left(0.295 + \frac{5\pi v}{2^{12}}\right) \sigma^{OBK}(1s - 1s). \tag{4.39}$$

Hence, asymptotically, there appears a term $\sigma \propto v^{-11}$ as in the Thomas-expression (4.37), but the coefficient is very small. The remarkable feature, which has attracted so much attention, is the fact that for very high speeds, the second order dominates the first. Other capture theories, like CDW, when carried to the second order, and SPB show similar results.

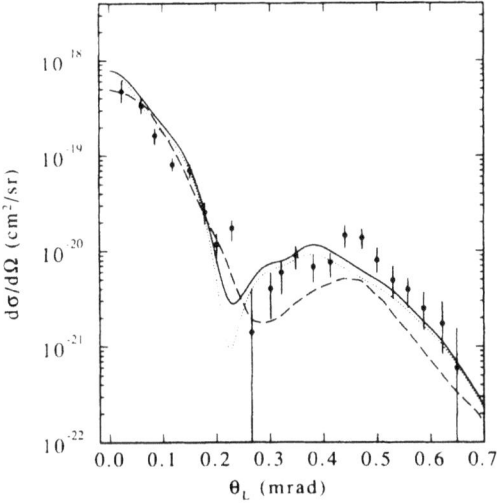

Figure 4.6: Differential cross sections for electron capture in $H^+ + H$ collisions at 5 Mev. The coupled-channel calculations [42] are performed with a contracted (physical) basis set of 34s, 57p and 30d states at the target and 34s states at the projectile. The physical s, p, and d basis states are expanded on 85, 75, and 66 GTO, respectively. Solid line: capture into the 1s, 2s, and 3s projectile states; dotted line: capture into 1s states only; dashed line: second-order boundary-corrected Born approximation (B2B0) [85]. Experimental data are from [86]. From [42].

In a large number of investigations using second-order theories, it has been shown that the sharp Thomas angle of classical mechanics is replaced by a broad peak in quantum mechanics, see Fig. 4.6.

The Thomas peak is also amenable to a coupled-channel calculation, see Sec. 4.7, although, for numerical reasons, nonperturbative approaches are mostly confined to moderate projectile velocities. By using Gaussian-type orbitals (GTO) (2.60), see Sec. 2.10, and (4.36) as a basis set in cartesian coordinates, it is possible to calculate matrix elements analytically and hence carry the treatment to very high collision velocities, for which the necessary number of matrix elements and the rapid oscillations of the translation factor usually sets a limit to the numerical evaluation [42].

The treatment breaks down in the relativistic energy regime.

Part II

Relativistic collisions

Chapter 5

Relativistic kinematics and fields of moving charges

It is the aim of this Chapter to provide a brief reminder of relativistic kinematics and to analyze the fields of moving charges. We start in Sec. 5.1 by introducing Lorentz transformations and by discussing their properties to the extent that is needed for the present purpose. In Sec. 5.2, we give explicit formulas for the transformation of energies and momenta between two inertial frames and in Sec. 5.3 furnish the expression for transforming differential cross sections. Subsequently, we study the relativistic motion of interacting point charges and the resulting deflection. The Liénard-Wiechert potentials produced by relativistically moving point charges are derived in Sec. 5.5, and the ensuing fields are represented as transient pulses of electromagnetic radiation. This leads in Sec. 5.6 to the equivalent-photon method, which is widely applied for estimating cross sections at very high relativistic energies.

5.1   The Lorentz transformation

The fundamental experimental fact that stands at the beginning of the Theory of Special Relativity is the observation by Michelson and Morley that the speed of light c is independent of the frame of reference, that is

$$c^2t^2 - \mathbf{x}^2 = c^2t'^2 - \mathbf{x}'^2 = 0. \tag{5.1}$$

Here, $(\mathbf{x}, t)$ and $(\mathbf{x}', t')$ are the space and time coordinates of a spherical shell of electromagnetic radiation observed in two different coordinate frames that move with a constant velocity v with respect to each other and coincide at $t = t' = 0$. The transition between the two frames is mediated by a "Lorentz transformation". The simultaneous transformation of space and time coordinates $\mathbf{x} = (x, y, z)$ and t, subject to Eq. (5.1), suggests that one can combine time and space coordinates into a four-vector

$$x^\mu = (x^0, x^1, x^2, x^3) = (ct, x, y, z) = (ct, \mathbf{x}). \tag{5.2}$$

In order to account for the minus sign in the invariant quadratic form (5.1), it is customary [51, 87, 88, 89] to introduce two different forms of a given four-vector, namely the *contravariant* form $a^\mu = (a^0, a^1, a^2, a^3)$ and the *covariant* form $a_\mu = (a_0, a_1, a_2, a_3)$. Both are related by $a_0 = a^0$, $a_k = -a^k$ for $k = 1, 2, 3$, so that

$$
\begin{aligned}
a^\mu &= (a^0, \mathbf{a}) & \text{is a contravariant vector,} \\
a_\mu &= (a^0, -\mathbf{a}) & \text{is a covariant vector.}
\end{aligned}
\tag{5.3}
$$

One can rewrite this definition with the aid of a metric tensor $g_{\mu\nu}$ in the form

$$a_\mu = \sum_{\nu=0}^{3} g_{\mu\nu} a^\nu \equiv g_{\mu\nu} a^\nu, \tag{5.4}$$

where the quantities $g_{\mu\nu}$ are elements of the matrix

$$g_{\mu\nu} = g^{\mu\nu} = \begin{pmatrix} 1 & 0 & 0 & 0 \\ 0 & -1 & 0 & 0 \\ 0 & 0 & -1 & 0 \\ 0 & 0 & 0 & -1 \end{pmatrix}. \tag{5.5}$$

In Eq. (5.4) and in the following, we use the summation convention which defines that the occurrence of repeated Greek-letter indices implies a summation over the labels 0,1,2,3.

The metric of Eq. (5.5) is introduced with the aim that the invariant quadratic form of Eq. (5.1) can be interpreted as a scalar product. More generally, invariant quantities can be obtained as scalar products of two four-vectors defined as

$$a_\mu b^\mu = a^\mu b_\mu = a^0 b^0 - \mathbf{a} \cdot \mathbf{b}. \tag{5.6}$$

Another convention [90] to secure the negative sign in Eq. (5.6) is to use a formally Euclidean metric $g'_{\mu\nu} = \delta_{\mu\nu}$ with an imaginary fourth component $a_4 = ia_0$. It avoids distinction between covariant and contravariant four-vectors, so that $a_\mu = a^\mu = (\mathbf{a}, ia_0)$. This convention turns out to be more "fool-proof" in actual calculations, because one does not have to worry about contravariant or covariant vectors.

The space-time derivatives corresponding to Eq. (5.2) deserve particular attention. From the rules of implicit differentiation it follows that

$$\partial^\mu \equiv \frac{\partial}{\partial x_\mu} = \left(\frac{\partial}{\partial(ct)}, -\nabla \right) \qquad \text{contravariant vector}$$

$$\partial_\mu \equiv \frac{\partial}{\partial x^\mu} = \left(\frac{\partial}{\partial(ct)}, \nabla \right) \qquad \text{covariant vector} \qquad (5.7)$$

which is in contrast to Eq. (5.3) regarding the signs of the space components.

When considering collisions with relativistic velocities, there is no need to study general Lorentz transformations, it is sufficient to confine oneself to a "Lorentz boost", that is, a Lorentz transformation in one direction, usually chosen to coincide with the z-direction. We then may write explicitly

$$\begin{pmatrix} a'_0 \\ a'_x \\ a'_y \\ a'_z \end{pmatrix} = \begin{pmatrix} \gamma & 0 & 0 & -\beta\gamma \\ 0 & 1 & 0 & 0 \\ 0 & 0 & 1 & 0 \\ -\beta\gamma & 0 & 0 & \gamma \end{pmatrix} \begin{pmatrix} a_0 \\ a_x \\ a_y \\ a_z \end{pmatrix}, \qquad (5.8)$$

where we have introduced the reduced velocity β and the Lorentz factor γ as

$$\beta = \frac{v}{c} \qquad \text{and} \qquad \gamma = \frac{1}{\sqrt{1 - \beta^2}}. \qquad (5.9)$$

The inverse transformation is obtained from Eq. (5.8) by interchanging primed and unprimed four-vectors and by replacing β with $-\beta$. In both cases, the transformation between the 0- and the z-coordinate may be viewed as a rotation in the 0-z-plane by an imaginary angle $i\chi$. The quantity χ is usually denoted as "rapidity." Since the angles χ_1 and χ_2 in successive two-dimensional rotations add up algebraically, so that $\chi_3 = \chi_1 + \chi_2$, this is also true for the rapidities of successive boosts in the same direction of space. Using the relations

$$\beta = \tanh \chi, \qquad \gamma = \cosh \chi, \qquad \beta\gamma = \sinh \chi, \qquad (5.10)$$

the additivity of the rapidities in conjunction with the addition theorem of hyperbolic tangents yields the addition theorem of relativistic velocities

$$\beta_3 = \frac{\beta_1 + \beta_2}{1 + \beta_1 \beta_2}. \qquad (5.11)$$

As a special case of Eq. (5.3), one can construct a four-vector by combining the momentum vector $\mathbf{p} = (p_x, p_y, p_z)$ of the particle with its energy

E (including the rest energy mc^2) to $p^\mu = (E, \mathbf{p}c)$ [87]. By observing that it must be possible to express all physical laws in terms of Lorentz invariants, we can form just one invariant quantity from energy and momentum of a particle, namely

$$p_\mu p^\mu = E^2 - p^2 c^2 = m^2 c^4, \qquad (5.12)$$

which establishes the well-known relationship between energy and momentum. Lorentz invariance tells us that a physical quantity, here the mass m of a particle, is independent of the inertial coordinate frame or "Lorentz frame" in which it is measured. Besides the rest mass, there are no further Lorentz invariants for a single point particle.

Decomposing the energy E of a particle into the rest energy mc^2 and the kinetic energy T with $E = mc^2 + T$, we have the following useful relations

$$
\begin{aligned}
\beta &= \frac{pc}{E} \\
\gamma &= \frac{E}{mc^2} \\
\beta\gamma &= \frac{p}{mc} \\
\gamma - 1 &= \frac{T}{mc^2}.
\end{aligned}
\qquad (5.13)
$$

5.2   Transformation between a moving frame and the laboratory frame

While theoretical results are often obtained in a moving coordinate system, for example in the center-of-mass system or in the projectile system, measurements are always performed in the laboratory system. Therefore, the transformation between the two Lorentz frames is needed. We assume that the reduced velocity β and the Lorentz factor γ of the moving frame with respect to the laboratory frame are known. Suppose, in the moving coordinate system, a particle has a momentum p', with a direction determined by the polar angle θ' with respect to the z-axis, and total energy E'. One may then use Eq. (5.8) to transform these quantities into the laboratory system (unprimed quantities) according to the relations

$$
\begin{aligned}
p \sin\theta &= p' \sin\theta' \\
p \cos\theta &= \gamma(p' \cos\theta' + \beta E'/c) \\
E/c &= \gamma(E'/c + \beta p' \cos\theta')
\end{aligned}
\qquad (5.14)
$$

or, conversely,

$$\begin{aligned}
p' \sin \theta' &= p \sin \theta \\
p' \cos \theta' &= \gamma(p \cos \theta - \beta E/c) \\
E'/c &= \gamma(E/c - \beta p \cos \theta).
\end{aligned} \tag{5.15}$$

In most accelerators today, one has a fixed target and an impinging beam of particles with a velocity v defining a fixed-target Lorentz factor γ_{FT} according to Eq. (5.9). However, for a collider ring (e.g., the Relativistic Heavy Ion Collider, RHIC, at Brookhaven) with counterpropagating beams of like particles with equal and opposite velocities characterized by the Lorentz factor γ_{coll}, it follows from Eqs. (5.14) and (5.15) that the corresponding fixed-target gamma is

$$\gamma_{\mathrm{FT}} = 2\gamma_{\mathrm{coll}}^2 - 1. \tag{5.16}$$

This means that in order to achieve very high values of γ, it is much more efficient to have two beams of equal and opposite velocities collide rather than letting a single beam with the Lorentz factor γ_{FT} impinge on a fixed target of equal-mass atoms. In the former way, no energy is wasted into the center-of-mass motion. Therefore, one can achieve extremely high-energy collisions with rather low-energy machines. On the other hand, a colliding-beam experiment will have to cope with a reduced intensity as compared to a fixed-target experiment.

5.3 Transformation of differential cross sections

Let us consider physical quantities in different Lorentz systems, a primed one and an unprimed one. Total cross sections are invariant quantities, $\sigma' = \sigma$, because they can be obtained by a mere counting of events, an operation that must be the same in all Lorentz frames. Differential cross sections are transformed from one Lorentz frame to another by requiring that the number of particles passing through a differential solid angle in one frame be equal to the number of particles passing through the *corresponding* differential solid angle in the other frame. The important quantity is the ratio of the solid angles $|d\Omega'/d\Omega| = |d\cos\theta'/d\cos\theta|$ which can be calculated [91] from the Lorentz transformations Eqs. (5.14) and (5.15), see also [3]. The cross sections are the related by

$$\frac{d\sigma(\theta,\varphi)}{d\Omega} = \left| \frac{d\Omega'}{d\Omega} \right| \frac{d\sigma'(\theta',\varphi')}{d\Omega'}. \tag{5.17}$$

As a special case, we now consider the emission of electromagnetic radiation from a moving system with the reduced velocity β and the associated

Lorentz factor γ. Since electrodynamics is Lorentz invariant, the phase $\omega t - \mathbf{k} \cdot \mathbf{r}$ of an electromagnetic wave is an invariant quantity, the wave vector $\mathbf{k}$ and the frequency ω form a four-vector $k^{\mu} = (\omega/c, \mathbf{k})$ which transforms according to Eq. (5.8) and is characterized by the invariant scalar product $k^{\mu} k_{\mu} = 0$. Photons satisfying this relation, are sometimes denoted as "real" photons and are said to be "on shell".

Let us denote quantities referring to the moving system by primed symbols and quantities relating to the laboratory system by unprimed symbols. In analogy to Eq. (5.15), we then derive the Doppler shift formulas

$$
\begin{aligned}
k' \sin \theta' &= k \sin \theta \\
k' \cos \theta' &= \gamma k (\cos \theta - \beta) \\
\omega' &= \gamma \omega (1 - \beta \cos \theta)
\end{aligned}
\tag{5.18}
$$

where θ and θ' are the angles of $\mathbf{k}$ and $\mathbf{k}'$ with respect to the direction of the velocity $\mathbf{v}$. The angles are related by

$$
\cos \theta' = \frac{\cos \theta - \beta}{1 - \beta \cos \theta}.
\tag{5.19}
$$

The inverse equations are obtained by interchanging primed and unprimed quantities and reversing the sign of β. The elements of the solid angle entering into the differential cross section are related by

$$
\frac{d\Omega'}{d\Omega} = \frac{1}{\gamma^2 (1 - \beta \cos \theta)^2}.
\tag{5.20}
$$

The last equation in Eqs. (5.18) shows that in relativistic kinematics there is a *transverse* Doppler shift which, owing to the presence of the factor γ, leads to a frequency change even when $\theta = \pi/2$.

Having transformed the angles, we are now in a position to transform differential cross sections. Suppose we know the differential cross section $d\sigma'(\theta', \varphi')/d\Omega'$ in a moving (primed) system at a *fixed frequency* ω'. From Eqs. (5.17) and (5.20) we then obtain the cross section in the (unprimed) laboratory system in the form

$$
\frac{d\sigma(\theta, \varphi)}{d\Omega} = \frac{1}{\gamma^2 (1 - \beta \cos \theta)^2} \frac{d\sigma'(\theta', \varphi')}{d\Omega'},
\tag{5.21}
$$

where $\varphi' = \varphi$ and θ' is expressed by θ using Eq. (5.19). This formula may be applied for transforming the differential cross section for the emission of x-rays in radiative electron capture (REC) from the projectile system to the laboratory frame, see Chapter 10.

5.4 Relativistic motion of interacting point charges

In the preceding Sections, we have discussed Lorentz transformations between inertial frames and have applied them to the transformation of differential cross sections. We now have to turn to a more specific problem, namely the relativistic motion of two interacting point charges. The classical treatment given in Sec. 1.4 can be generalized to relativistic velocities, see, e.g. [92]. For the purely repulsive interaction between two positive point charges in which we are interested, deflection angle Θ, see Eq. (1.12) and scattering angle θ are identical and, within classical mechanics, there is a one-to-one relationship between the impact parameter and the scattering angle, for example Eq. (1.13). Small impact parameters correspond to large scattering angles and vice versa. However, as is known from Rutherford's classic scattering experiment for projectile energies as low as about 1 MeV/u, large deflection angles of the order of $\pi/2$ correspond to impact parameters of the order a few fm ($= 10^{-13}$cm). On the other hand, the length scale for atomic processes is given by the atomic K-shell radius $a_{\mathrm{K}} = Z^{-1}a_0 = Z^{-1} \times 0.53 \times 10^{-8}$cm, see Eq. (1.3), where Z is the nuclear charge number involved. Since the contribution to the cross section of an impact parameter range between b and $b + db$ is weighted with $2\pi\,b\,db$, we expect that impact parameters of the order of a few fm, which are associated with large deflection angles, give a negligible contribution. Moreover, for atomic collision studies, these small impact parameters have to be avoided because the background from nuclear reactions would tend to mask atomic processes. Therefore, it is perfectly justified for most purposes to disregard the finite nuclear size and to substitute a positive point charge for the nucleus.

It is often convenient to express the Lorentz factor γ directly by the kinetic energy of the projectile as

$$\gamma = \frac{1}{\sqrt{1 - \beta^2}} = 1 + \frac{T_{\mathrm{P}}\,(\mathrm{MeV/u})}{931.494} \tag{5.22}$$

and to write the deflection angle, see Fig. 1.1, as a function of the ratio

$$x = \frac{e^2 Z_{\mathrm{P}} Z_{\mathrm{T}}/b}{M_{\mathrm{P}} c^2} \tag{5.23}$$

between the Coulomb repulsion energy of the nuclei at the separation b and the rest energy of the projectile. Using the abbreviation

$$\eta = \sqrt{1 - \frac{x^2}{\gamma^2 - 1}}, \tag{5.24}$$

one obtains the deflection angle Θ in the center-of-mass system [92] as a function of the parameters γ and x as

$$\Theta = \theta = \pi - \frac{2}{\eta} \arctan \frac{(\gamma^2 - 1)\eta}{\gamma x}, \qquad (5.25)$$

or, for $x \ll 1$ and $\gamma \gg 1$,

$$\theta \approx 2 \frac{x}{\gamma}. \qquad (5.26)$$

Let us consider U + U collisions as an example. It is seen that down to impact parameters b very small on the atomic scale, the Coulomb deflection is negligible for relativistic collisions. Even for $b = 15$ fm, when the uranium nuclei begin to touch each other, and for a kinetic energy of 1 GeV/u in the cm system, the more accurate estimate using Eq. (5.25) yields a deflection of only 9 mrad. As a result, it is seen that it is perfectly legitimate in cases of interest to use classical straight-line trajectories in a semiclassical description, in which only the electron motion is treated within quantum theory.

5.5   Liénard-Wiechert potentials

As a consequence of the estimates given in Sec. 5.4, we assume in the following that the projectile moves along a classical rectilinear trajectory given by $\mathbf{R} = \mathbf{b} + \mathbf{v}t$, where $\mathbf{b}$ is the impact parameter vector and $\mathbf{v}$ is a constant velocity. When defining the coordinate systems, it is convenient to place the target nucleus at the origin of the laboratory system with the x- and z-axes taken in the directions of $\mathbf{b}$ and $\mathbf{v}$, respectively. The inertial frame, in which the projectile nucleus is at rest, moves with the speed v with respect to the laboratory frame, and its x', y', z'-axes are parallel to the x, y, z-laboratory axes, respectively. For the time coordinates, we choose the convention that the times t_0, t_0' associated with the nuclear positions are both zero when the projectile reaches its closest approach to the target nucleus during the collision. Let $x^\mu = (ct, x, y, z)$ and $x'^\mu = (ct', x', y', z')$ denote the space-time coordinates of a point with respect to the two Lorentz frames. With the Lorentz transformation (5.8), taking into account the displacement by the impact parameter $\mathbf{b}$, and defining the unit vector $\hat{\mathbf{v}} = \mathbf{v}/v$, the connection between the two coordinate systems can be written in vector form as

$$\begin{aligned} \mathbf{r}' &= \mathbf{r} + (\gamma - 1)(\mathbf{r} \cdot \hat{\mathbf{v}})\hat{\mathbf{v}} - \gamma \mathbf{v}t - \mathbf{b} \\ t' &= \gamma(t - \mathbf{r} \cdot \mathbf{v}/c^2), \end{aligned} \qquad (5.27)$$

or, for the inverse transformation,

$$\mathbf{r} = \mathbf{r'} + (\gamma - 1)(\mathbf{r'} \cdot \hat{\mathbf{v}})\hat{\mathbf{v}} + \gamma \mathbf{v} t' + \mathbf{b}$$
$$t = \gamma(t' + \mathbf{r'} \cdot \mathbf{v}/c^2). \tag{5.28}$$

Choosing the z-axis in the direction of $\mathbf{v}$, we have explicitly

$$\begin{pmatrix} ct' \\ x' \\ y' \\ z' \end{pmatrix} = \begin{pmatrix} \gamma & 0 & 0 & -\beta\gamma \\ 0 & 1 & 0 & 0 \\ 0 & 0 & 1 & 0 \\ -\beta\gamma & 0 & 0 & \gamma \end{pmatrix} \begin{pmatrix} ct \\ x \\ y \\ z \end{pmatrix} - \begin{pmatrix} 0 \\ b \\ 0 \\ 0 \end{pmatrix}, \tag{5.29}$$

and the distance of a point (x, y, z, t) from the origin of the moving frame, measured in the moving frame, is

$$r' = \sqrt{(x - b)^2 + y^2 + \gamma^2(z - vt)^2}. \tag{5.30}$$

The same transformations allow us to construct the four-potential $A^\mu = (A^0, \mathbf{A}) = (\Phi, \mathbf{A})$ generated by a uniformly moving charge. In the projectile frame, the projectile charge eZ_P gives rise to the static potentials

$$\Phi'(\mathbf{r'}, t') = \frac{eZ_\mathrm{P}}{r'}, \qquad \mathbf{A'}(\mathbf{r'}, t') = 0. \tag{5.31}$$

By the defining relation of a four-vector, namely that it transforms exactly like the space-time coordinates, the corresponding four-potentials Φ and $\mathbf{A}$ in the target frame are obtained by the homogeneous part of the same Lorentz transformation (5.29) which carries the four-coordinates t' and $\mathbf{r'}$ into t and $\mathbf{r}$. By simple substitution, we derive the Liénard-Wiechert potentials produced by the projectile in the target frame in the forms

$$\Phi(\mathbf{r}, t) = \frac{\gamma e Z_\mathrm{P}}{\sqrt{(x - b)^2 + y^2 + \gamma^2(z - vt)^2}}$$

$$\mathbf{A}(\mathbf{r}, t) = \frac{\mathbf{v}}{c} \Phi(\mathbf{r}, t), \tag{5.32}$$

which satisfy the Lorentz gauge condition

$$\frac{1}{c} \frac{\partial \Phi}{\partial t} + \nabla \cdot \mathbf{A} = 0. \tag{5.33}$$

In particular, at the origin of the target frame, the electric and magnetic fields due to the moving projectile are given by, see [87],

$$
\begin{aligned}
E_x &= \frac{e Z_{\mathrm{P}}\,\gamma b}{(b^2 + \gamma^2 v^2 t^2)^{3/2}} \\
E_y &= 0 \\
E_z &= -\frac{e Z_{\mathrm{P}}\,\gamma v t}{(b^2 + \gamma^2 v^2 t^2)^{3/2}}
\end{aligned}
\tag{5.34}
$$

and

$$
\begin{aligned}
B_x &= 0 \\
B_y &= \beta\, E_x \\
B_z &= 0,
\end{aligned}
\tag{5.35}
$$

respectively. We see that the peak transverse electric field E_x is increased

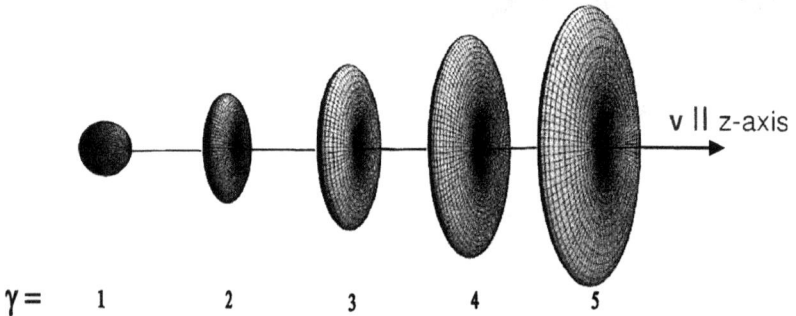

Figure 5.1: Polar diagrams for the angular dependence of the radial electric field strength produced by a point charge moving with the velocity $\mathbf{v}$ to the right. The numbers give the Lorentz factors γ. From [93].

by a factor of γ with respect to its nonrelativistic value, while the duration

$$
\Delta t \simeq b/(\gamma v)
\tag{5.36}
$$

of appreciable field strength at the target nucleus is decreased by the same factor. We note that the ratio $E_x/E_z = -b/(vt)$ is just the tangent of the angle $\theta = \arccos(-\hat{\mathbf{R}} \cdot \hat{\mathbf{v}})$ formed between the vectors $-\mathbf{R}$ and the z-axis (Fig. 1.1). Hence the electric field produced by the charge $e Z_{\mathrm{P}}$ at the position of the target nucleus is directed radially from the projectile's present (i.e., not retarded) position to the observation point at the target nucleus. Writing $b = R \sin\theta$ and $vt = R\cos\theta$, we obtain the electric field

$$\mathbf{E} = \frac{-eZ_P\,\mathbf{R}}{\gamma^2 R^3 (1 - \beta^2 \sin^2\theta)^{3/2}}. \tag{5.37}$$

The angular variation of the radial electric field strength is illustrated in Fig. 5.1 for various projectile velocities in terms of the Lorentz factor γ. Along the direction of motion, $\theta = 0$ or $\theta = \pi$, the field strength is decreased by a factor of γ^{-2} as compared to a charge at rest. On the other hand, perpendicular to the trajectory, $\theta = \pi/2$, the field is increased by a factor of γ. The flattening of the surfaces into disk shapes with a depression in the center is essentially an effect of the Lorentz contraction of the electromagnetic fields, but clearly does not follow the Lorentz contraction of a shere in space.

5.6 The equivalent-photon method

The electric field produced by a moving charge becomes almost transverse if $\gamma \gg 1$, see Eqs. (5.34) and (5.35), and is accompanied by a transverse magnetic field perpendicular to it and of almost equal strength. These electric and magnetic fields can be replaced approximately by the fields of a pulse of plane linearly polarized radiation propagating in the z-direction. This replacement forms the basis of the equivalent-photon method which is originally due to Fermi [94] and was developed by von Weizsäcker [95] and Williams [96]. The method can be applied to a variety of electromagnetic processes induced by charged projectiles in nuclei and atoms. These applications have been reviewed in [97] and, more recently in [98], see also [3]. The Weizsäcker-Williams or equivalent-photon method is valuable at extremely high energies in situations where rigorous methods are difficult to implement. It is then possible to use the known cross sections for photon-induced reactions to estimate the corresponding cross sections induced by moving charges.

The main idea behind the equivalent-photon method is to replace the transient fields (5.34) and (5.35) by two pulses, P_1 and P_2 of linearly polarized radiation. Figure 5.2 illustrates the photon spectrum of the two pulses as a function of a reduced frequency $\zeta = \omega b/\gamma v$. For larger Lorentz factors, for which the method becomes applicable, the pulse P_2 is completely neglegible, so that the collision is effectively represented by the photon spectrum of the pulse P_1, whose decrease around $\zeta = 1$ becomes steeper with increasing collision energy. However, while for real photons one has $k = \omega/c$, this is not so for the equivalent or virtual photons, which are

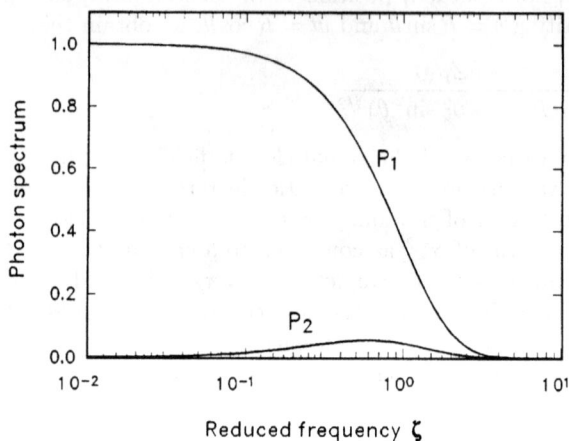

Figure 5.2: The shape of the equivalent-photon spectrum for the Pulses P_1 and P_2 as a function of $\zeta = \omega b/\gamma v$ and assuming a Lorentz factor $\gamma = 2$. The cut-off frequency (5.38) corresponds to $\zeta \approx 1$. From [3].

off the energy shell and only approach the energy shell $k_\mu k^\mu = 0$ for very high energies. According to Eq. (5.36), the time during which the fields are appreciable is of the order

$$\Delta t \approx \frac{b}{\gamma v} \, . \tag{5.38}$$

Note that for large values of γ, this defines a considerable reduction compared to the nonrelativistic estimate of the collision time as b/v. During this time interval, the fields are $|E_z| = |E_x|vt/b \approx |E_x|/\gamma$.

In applying the equivalent-photon approximation, we consider a swarm of real (on-shell) photons for which the relation $k = \omega/c$ between the wave number k and the frequency ω holds. The task then consists in decomposing the classical fields into photon spectra, an example of which is given in Fig. 5.2, and to calculate the number $N(\omega, b)$ of equivalent photons with a given frequency ω produced by the passage of a projectile at the impact parameter b, see e.g., [3, 87, 97]. The known photon-induced cross sections

Table 5.1: For the heavy-ion accelerators GSI (Darmstadt) with its future extension, the AGS (Brookhaven) and the colliders RHIC (Brookhaven), and LHC (Geneva), typical projectile energies E are given (for colliders in the collider frame). The resulting maximum photon energies $\hbar\omega_{max}$ for the impact parameter $b = 386$ fm (equal to the electron Compton wave length) and $b = 7.7$ fm (corresponding to grazing collisions between heavy nuclei) are listed in the last two columns.

	E (GeV/u)	γ	$\hbar\omega_{max}$ at λ_c (MeV)	$\hbar\omega_{max}$ at 7.7 fm (MeV)
GSI	1	2.1	1.1	53
	5	6.4	3.3	163
	10	11.7	6.0	300
	30	33.2	17.0	850
AGS	12	13.9	7.2	370
RHIC	100	108.3	60	3 000
LHC	3 400	3 651	2 000	100 000

multiplied by $N(\omega, b)$ then yield estimates for cross sections arising from relativistically moving charges. Since from Eq. (5.38) the duration of the passage is given by $\Delta t \simeq b/\gamma v$, frequencies appreciably higher than

$$\omega_{max} \approx \frac{1}{\Delta t} \approx \frac{\gamma c}{b} \tag{5.39}$$

cannot occur. In Table 5.1, we present maximum photon energies

$$\hbar\omega_{max} \approx 0.511\, \gamma\, \frac{\lambda_c}{b} \ \text{(MeV)} \tag{5.40}$$

corresponding to projectile energies representative for some heavy-ion accelerators. With regard to the equivalent-photon method, one has to be aware of the limitations arising from the assumption of real photons and from the assumption of a constant field across the transverse extension of the perturbed system. The latter restriction may invalidate the method for impact parameters smaller than the Compton wavelength of the electron.

Chapter 6

Relativistic electron motion

In Chapter 5 we have treated the relativistic motion of the colliding heavy particles, the ions and the atoms. While it is well justified to apply classical relativistic mechanics in this case, the electron motion requires a quantum desription. Furthermore, in a high-Z atom, the electrons in initial or final states are subject to a strong Coulomb potential demanding a relativistic description. The Bohr velocity of an electron in a $1s_{1/2}$ orbital is $\alpha Z\,c \approx (Z/137)\,c$ and hence comparable to the speed c of light.

In this Chapter, we summarize properties of the Dirac equation for a central potential. In Sec. 6.1, we introduce the appropriate quantum numbers, and in Sec. 6.2 proceed to a classification of bound states in a Coulomb potential and provide the corresponding wavefunctions. Finally, in Sec. 6.3, we present Coulomb-Dirac continuum wavefunctions, both as a partial-wave expansion and in the approximate Sommerfeld-Maue form.

6.1 The Dirac equation for a central potential

A quantum treatment of the electron motion requires the use of the Dirac equation, which is the relativistic wave equation for a spin-$\frac{1}{2}$ particle. Within this theory, the wavefunction is given by a four-component spinor in which the upper two components, with spin up and spin down, refer to positive-energy states while the lower two components refer to the negative-energy states. For easy reference and for establishing the notation, we here summarize some of the basic properties of the Dirac equation and its solutions. Writing the time-dependent wavefunction for a spherical potential $V(r)$ quite generally in the form

$$\psi(\mathbf{r}, t) = \varphi(\mathbf{r})\, e^{-\mathrm{i}Et/\hbar}, \tag{6.1}$$

the eigenstates are defined as solutions of the stationary Dirac equation

$$H\varphi(\mathbf{r}) = \left(-i\hbar c\boldsymbol{\alpha} \cdot \nabla + V(r) + m_e c^2 \gamma^0\right) \varphi(\mathbf{r}) = E\varphi(\mathbf{r}), \qquad (6.2)$$

where $\mathbf{r}$ is the coordinate of the electron with respect to the nucleus, $\boldsymbol{\alpha}$ denotes the vector formed from the usual 4×4 Dirac matrices, γ^0 is the fourth matrix, often denoted by β. These matrices, as well as the 4×4 spin matrix $\boldsymbol{\sigma}_4$, are constructed from the 2×2 Pauli matrices $\boldsymbol{\sigma}$ and the 2×2 unit matrix $\mathbf{I}_2$ as

$$\boldsymbol{\alpha} = \begin{pmatrix} 0 & \boldsymbol{\sigma} \\ \boldsymbol{\sigma} & 0 \end{pmatrix}, \qquad \gamma^0 = \begin{pmatrix} \mathbf{I}_2 & 0 \\ 0 & -\mathbf{I}_2 \end{pmatrix}, \qquad \boldsymbol{\sigma}_4 = \begin{pmatrix} \boldsymbol{\sigma} & 0 \\ 0 & \boldsymbol{\sigma} \end{pmatrix}. \quad (6.3)$$

The quantity E in (6.2) is the eigenenergy. The wavefunctions $\phi(\mathbf{r})$ are represented by 4-spinors (see, e.g., [90, 99]). Having defined the Dirac equation, we will adopt relativistic units $\hbar = m_e = c = 1$ in this Chapter and will only expose the full units explicitly, when it serves the physical understanding.

For a relativistic electron, the angular-momentum operator $\mathbf{J} = \mathbf{L} + \mathbf{s}$, where $\mathbf{L}$ is the orbital angular momentum operator and $\mathbf{s} = \frac{1}{2}\boldsymbol{\sigma}_4$, is the 4×4 spin operator. Since $\mathbf{J}^2$ as well as J_z commute with the Hamiltonian H contained in Eq. (6.2), one can construct simultaneous eigenfunctions of $H, \mathbf{J}^2, J_z$ and parity. The eigenfunctions are labelled by the Dirac quantum number κ combining j and parity:

$$\kappa = \mp(j + \tfrac{1}{2}) \quad \text{as} \quad j = l \pm \tfrac{1}{2} \quad (\text{or} \quad j = l' \mp \tfrac{1}{2}). \qquad (6.4)$$

Here l is the orbital angular momentum quantum number of the *upper components* of the four-spinor (which is usually used for the spectroscopic notation), while l' is the orbital momentum quantum number of the *lower components*. This means that the Dirac quantum numbers $\kappa = -1, +1, -2, +2, -3, \cdots$ are assigned to the spectroscopic notations $s_{1/2}, p_{1/2}, p_{3/2}, d_{3/2}, d_{5/2}, \cdots$ states, see also Table 6.1

Introducing the normalized spin-angular functions $\chi_\kappa^{m_j}$ defined as eigenfunctions of $\mathbf{J}^2, \mathbf{L}^2, J_z$, characterized by the angular momentum j and the projection m_j as

$$\chi_\kappa^{m_j}(\hat{\mathbf{r}}) = \sum_{m_l} \left(\begin{array}{cc|c} l & \frac{1}{2} & j \\ m_l & m_j - m_l & m_j \end{array} \right) Y_{lm_l}(\hat{\mathbf{r}}) \chi_{\frac{1}{2}}^{m_j - m_l}, \qquad (6.5)$$

where $(\ \ |\ )$ is a Clebsch-Gordan coefficient, see e.g., [100] and $\chi_{1/2}^{\pm 1/2}$ is a Pauli spinor, we can write the solutions of the Dirac equation (6.2) in the form

$$\varphi_{\kappa m_j}(\mathbf{r}) = \begin{pmatrix} g_\kappa(r)\chi_\kappa^{m_j}(\hat{\mathbf{r}}) \\ if_\kappa(r)\chi_{-\kappa}^{m_j}(\hat{\mathbf{r}}) \end{pmatrix}. \tag{6.6}$$

With the substitutions $F(r) = r\,f(r)$, $G(r) = r\,g(r)$, we finally obtain from Eq. (6.2) the coupled radial equations for a general spherical potential $V(r)$ as

$$\left(\frac{dG}{dr} + \frac{\kappa}{r}G\right) = (E - V + 1)F$$

$$\left(\frac{dF}{dr} - \frac{\kappa}{r}F\right) = -(E - V - 1)G. \tag{6.7}$$

By construction, the Dirac equation is Lorentz-invariant. In the following, we confine ourselves to those transformations in which we are primarily interested, namely Lorentz boosts in the z-direction. In this case, one finds that the spinor transformation

$$\psi'(\mathbf{x}') = S\,\psi(\mathbf{x}) \tag{6.8}$$

mediating between a moving coordinate (primed) system and a system at rest (unprimed), is achieved by the 4×4 matrix

$$S(\pm v) = S^\dagger(\pm v) = \sqrt{\frac{1+\gamma}{2}}\,(1 \mp \delta\alpha_z), \tag{6.9}$$

where the Dirac matrix α_z is the z-component of $\boldsymbol{\alpha}$, and

$$\delta = \sqrt{\frac{\gamma - 1}{\gamma + 1}} \tag{6.10}$$

is a parameter measuring the magnitude of the relativistic corrections arising from the Lorentz transformation. The matrix S can be shown [3] to have the property $\gamma^0 S \gamma^0 = S^{-1}$.

In the transformation (6.8) for Dirac spinors, the matrix $S = S(v)$ depends only on the relative velocity of the two coordinate frames with respect to one another. For later reference, we also give the explicit form of the square as

$$S^2(v) = S^{-2}(-v) = \gamma(1 - \beta\alpha_z) \tag{6.11}$$

with $\beta = v/c$, which is obtained from Eqs. (6.9) and (6.10) by direct matrix multiplication. Equation (6.11) and the relation $\gamma^0 S \gamma^0 = S^{-1}$ play an important role in the formulation of relativistic atomic collisions. While the matrix (6.9) transforms the relativistic wavefunction, the matrix (6.11) transforms the potentials.

6.2   Bound states in a Coulomb potential

We now turn to a hydrogen-like ion by specifying $V(r) = -e^2 Z/r$, and, introducing the abbreviations

$$s = \sqrt{\kappa^2 - \zeta^2} \quad \text{and} \quad \zeta = \alpha Z, \tag{6.12}$$

Table 6.1: Relativistic quantum numbers of hydrogen-like states and spectroscopic notation. Pairs of states degenerate according to Eq. (6.13) are grouped together. For the definition of l and l', see Eq. (6.4).

shell	n	$n' = n - \lvert\kappa\rvert$	$\kappa = \mp(j + \frac{1}{2})$	j	l	l'	notation
K	1	0	-1	$1/2$	0	1	$1\mathrm{s}_{1/2}$
L	2	1	-1	$1/2$	0	1	$2\mathrm{s}_{1/2}$
		1	$+1$	$1/2$	1	0	$2\mathrm{p}_{1/2}$
		0	-2	$3/2$	1	2	$2\mathrm{p}_{3/2}$
M	3	2	-1	$1/2$	0	1	$3\mathrm{s}_{1/2}$
		2	$+1$	$1/2$	1	0	$3\mathrm{p}_{1/2}$
		1	-2	$3/2$	1	2	$3\mathrm{p}_{3/2}$
		1	$+2$	$3/2$	2	1	$3\mathrm{d}_{3/2}$
		0	-3	$5/2$	2	3	$3\mathrm{d}_{5/2}$

we may write down the Sommerfeld formula for the hydrogenic eigenenergies as

$$E_{n\kappa} = \left[1 + \left(\frac{\zeta}{n' + s}\right)^2\right]^{-\frac{1}{2}}, \tag{6.13}$$

where $n' = 0, 1, 2, \cdots$ counts the nodes of the radial wavefunction and is related to the principal quantum number $n = 1, 2, 3, \cdots$ by $n = n' + |\kappa|$. Since $E_{n\kappa} < 1$, Eq. (6.13) can be expanded in terms of $\zeta = \alpha Z$ to give

$$E_{n\kappa} = 1 - \frac{1}{2}\frac{(\alpha Z)^2}{n^2} - \frac{1}{2}\frac{(\alpha Z)^4}{n^3}\left(\frac{1}{j + \frac{1}{2}} - \frac{3}{4n}\right) - \cdots, \tag{6.14}$$

where the first term is the rest mass, and the second term represents the nonrelativistic binding energy in a hydrogenic atom.

Now, suppressing the labels n and κ and introducing the bound-state wave number

$$q = \sqrt{1 - E_{n\kappa}^2} = \zeta\left[\zeta^2 + (n' + s)^2\right]^{-\frac{1}{2}}, \tag{6.15}$$

the radial functions $g(r)$ and $f(r)$ can be expressed by confluent hypergeometric functions $_1F_1(a, c; x)$ as

$$\begin{aligned}
g_\kappa(r) &= N_g (2qr)^{s-1} e^{-qr} \left[-n' \, _1F_1(-n' + 1, 2s + 1; 2qr)\right. \\
&\quad \left. -(\kappa - \frac{\zeta}{q}) \, _1F_1(-n', 2s + 1; 2qr)\right]
\end{aligned}$$

$$\begin{aligned}
f_\kappa(r) &= N_f (2qr)^{s-1} e^{-qr} \left[n' \, _1F_1(-n' + 1, 2s + 1; 2qr)\right. \\
&\quad \left. -(\kappa - \frac{\zeta}{q}) \, _1F_1(-n', 2s + 1; 2qr)\right]
\end{aligned} \tag{6.16}$$

with

$$N_g = \frac{\sqrt{2}\, q^{\frac{5}{2}}}{\Gamma(2s + 1)} \left[\frac{\Gamma(2s + n' + 1)(1 + E_{n\kappa})}{n'!\, \zeta(\zeta - \kappa q)}\right]^{\frac{1}{2}},$$

$$N_f = -N_g \left(\frac{1 - E_{n\kappa}}{1 + E_{n\kappa}}\right)^{\frac{1}{2}}. \tag{6.17}$$

Since for $|\kappa| = 1$ one has $s < 1$ according to Eq. (6.12), a mild singularity appears at the origin in the wavefunctions for $1s_{1/2}$ and, similarly, for $2p_{1/2}$ states.

When evaluating Eq. (6.13) for the energy of excited states, we find that within one principal shell n, the energy is the same for equal values

of j, but for equal values of l it is different. For a given value of l, the
spin-orbit splitting between states with $j = l + \frac{1}{2}$ and $j = l - \frac{1}{2}$ gives rise to
the fine structure in the spectrum of hydrogenlike atoms. Table 6.1 gives
the lowest hydrogenic states together with their spectroscopic notation.

For many purposes, it is sufficient to consider the hydrogenic ground
state $1s_{1/2}$ with $\kappa = -1$. Here, $E_{n\kappa} = s$ and, with the Bohr radius $a_0 = \lambdabar_c/\alpha = \hbar^2/m_e e^2$, we have

$$\varphi^{\frac{1}{2}}_{1s_{1/2}}(\mathbf{r}) = \frac{1}{\sqrt{4\pi}} \begin{pmatrix} g(r) \\ 0 \\ -if(r)\cos\theta \\ -if(r)\sin\theta e^{i\phi} \end{pmatrix},$$

$$\varphi^{-\frac{1}{2}}_{1s_{1/2}}(\mathbf{r}) = \frac{1}{\sqrt{4\pi}} \begin{pmatrix} 0 \\ g(r) \\ -if(r)\sin\theta e^{-i\phi} \\ if(r)\cos\theta \end{pmatrix} \tag{6.18}$$

with the radial wavefunctions

$$g(r) = N_g \left(\frac{r}{a_0}\right)^{s-1} e^{-Zr/a_0}$$

$$f(r) = N_f \left(\frac{r}{a_0}\right)^{s-1} e^{-Zr/a_0}, \tag{6.19}$$

where

$$N_g = \frac{(2Z)^{s+\frac{1}{2}}}{\left[2\,\Gamma(2s+1)\right]^{\frac{1}{2}}} (1+s)^{\frac{1}{2}}, \qquad N_f = -\left(\frac{1-s}{1+s}\right)^{\frac{1}{2}} N_g. \tag{6.20}$$

We are now in possession of exact relativistic wavefunctions for the
bound Coulomb problem. Nevertheless, it is often more convenient, and for
small values of $\zeta = \alpha Z$ also sufficient, to use approximate wavefunctions of a
considerably simpler structure [3, 99, 89]. Denoting with $u^{(+)} = (1,0,0,0)^\dagger$
and $u^{(-)} = (0,1,0,0)^\dagger$ the basic four-component spinors for a particle at
rest with spin up and spin down, respectively, and retaining terms up to
order αZ, one may derive, see e.g., [3], the Darwin wavefunction [101]

$$\varphi^{(\pm)}(\mathbf{r}) = \left(1 - \frac{i\hbar}{2m_e c}\boldsymbol{\alpha}\cdot\nabla\right) u^{(\pm)}\varphi_0(\mathbf{r}) \tag{6.21}$$

as a quasirelativistic bound-state wavefunction accurate to first order in
αZ in the relativistic correction (and normalized to the same order), with

φ_0 being a nonrelativistic bound-state hydrogenic function. The method is also applicable to more general potentials.

6.3 Coulomb-Dirac continuum wavefunctions

Similarly as in the case of bound-state wavefunctions, we may treat each partial wave, characterized by the quantum numbers κ and m_j, by solving the corresponding radial equations (6.7) with the appropriate boundary conditions. While for the calculation of total cross sections it is sufficient to know the individual partial waves and add their contributions incoherently, the study of differential cross sections requires the construction of complete scattering solutions and the knowledge of the appropriate phases. We start with the partial waves.

For a continuum electron $E > 1$, the corresponding momentum, see Eq. (5.12), is

$$p = \sqrt{E^2 - 1} \tag{6.22}$$

and the spinor wave function has the form corresponding to Eq. (6.6)

$$\varphi_{E,\kappa m_j}(\mathbf{r}) = \begin{pmatrix} g_\kappa(r)\chi_\kappa^{m_j}(\hat{\mathbf{r}}) \\ \mathrm{i}f_\kappa(r)\chi_{-\kappa}^{m_j}(\hat{\mathbf{r}}) \end{pmatrix}. \tag{6.23}$$

Introducing the relativistic generalization of the Sommerfeld parameter ν defined in Eq. (4.15) as

$$\eta = \frac{\zeta E}{p} = \frac{\alpha Z E}{p}, \tag{6.24}$$

a phase factor δ_κ, and a normalization factor N_κ by

$$\begin{aligned} \delta_\kappa &= \tfrac{1}{2}\arg\left(\frac{-\kappa + \mathrm{i}\eta/E}{s + \mathrm{i}\eta}\right) \\ N_\kappa &= \sqrt{\frac{2p}{\pi}}\,\mathrm{e}^{\pi\eta/2}\,\frac{|\Gamma(s + \mathrm{i}\eta)|}{\Gamma(2s + 1)}, \end{aligned} \tag{6.25}$$

we can write, as a relativistic extension of (3.30), the radial continuum wavefunctions entering in the spinor (6.23) for a given partial wave with $E > 1$ as

$$\begin{aligned} g_\kappa &= N_\kappa(E + 1)^{\frac{1}{2}}(2pr)^{s-1}\,\mathrm{Re}\left[\mathrm{e}^{-\mathrm{i}pr}\mathrm{e}^{\mathrm{i}\delta_\kappa}(s + \mathrm{i}\eta)\right. \\ &\qquad\qquad \left. \times\ {}_1F_1(s + 1 + \mathrm{i}\eta, 2s + 1; 2\mathrm{i}pr)\right], \end{aligned}$$

$$f_\kappa \;=\; -N_\kappa (E-1)^{\frac{1}{2}} (2pr)^{s-1} \, \mathrm{Im} \left[e^{-ipr} e^{i\delta_\kappa} (s+i\eta) \right.$$
$$\left. \times \;_1F_1(s+1+i\eta, 2s+1; 2ipr) \right]. \quad (6.26)$$

The wavefunctions (6.26) are normalized on the energy scale, see Eq. (3.27). This means that if φ_E and $\varphi_{E'}$ are solutions with eigenenergies E and E' we have

$$\int \varphi^\dagger_{E',\kappa m_j}(\mathbf{r}) \varphi_{E,\kappa m_j}(\mathbf{r}) \, \mathrm{d}^3 r = \delta(E-E'). \quad (6.27)$$

The asymptotic behavior of the wavefunctions (6.26), which has to be chosen consistently with the Coulomb phases, is given by

$$g_\kappa \;\simeq\; \frac{1}{r} \left(\frac{E+1}{\pi p} \right)^{\frac{1}{2}} \cos(pr + \sigma_\kappa),$$

$$f_\kappa \;\simeq\; -\frac{1}{r} \left(\frac{E-1}{\pi p} \right)^{\frac{1}{2}} \sin(pr + \sigma_\kappa) \quad (6.28)$$

Adopting the phase convention of Eq. (6.28), the Coulomb phase shift is

$$\sigma_\kappa = \delta_\kappa - \arg \Gamma(s+i\eta) - \tfrac{1}{2}\pi s + \eta \, \ln(2pr). \quad (6.29)$$

For expressing the scattering waves, we also introduce

$$\Delta_\kappa = \delta_\kappa - \arg \Gamma(s+i\eta) - \tfrac{1}{2}\pi s + (l+1)\tfrac{\pi}{2}. \quad (6.30)$$

Sometimes a phase convention for the radial wavefunctions is chosen which differs from the convention adopted in (6.28). Then the Coulomb phases (6.29) and (6.30) have to be modified accordingly[†].

If it is the goal to calculate total cross sections for electron emission, that is, if no information is needed regarding the direction of asymptotic propagation of the electron, it is sufficient to sum incoherently over all partial waves contributing to the cross section. If, however, the direction of propagation of an emitted electron is of interest, appropriate superpositions of partial waves must be used.

Introducing a partial-wave expansion into states characterized by $\kappa = (l,j)$ and μ, it is convenient for calculating total cross sections and necessary for angular correlations and spin effects, see, e.g., [3], to employ the helicity representation of the electron continuum wavefunction, in which

---

[†]Note that in [3] there is an inconsistency in phases. In Eqs. (4.120) and (9.39) a phase $(l+1)\tfrac{\pi}{2}$ should be added to Δ_κ and, correspondingly, subtracted from σ_κ in Eqs. (4.120) and (9.38).

the electron spin is quantized along the asymptotic momentum. In fact, a relativistic electron has a sharp value of its spin projection m_s only with respect to its own direction of motion. Adopting the asymptotic normalization to unit-amplitude plane waves, the function

$$
\psi_{\mathbf{p},m_s}(\mathbf{r}) \;=\; \sum_{\kappa,\mu} i^l e^{-i\Delta_\kappa}\sqrt{4\pi(2l+1)}\left(\begin{array}{cc|c} l & \tfrac{1}{2} & j \\ 0 & m_s & m_s \end{array}\right)
$$

$$
\times \left(\begin{array}{c} g_\kappa(r)\,\chi_\kappa^\mu \\ if_\kappa(r)\,\chi_{-\kappa}^\mu \end{array}\right) D_{\mu m_s}^j(\hat{\mathbf{z}} \to \hat{\mathbf{p}}) \qquad (6.31)
$$

is the partial-wave expansion with the spin projection m_s with respect to $\mathbf{p}$. The Wigner rotation matrix $D_{\mu m_s}^j(\hat{\mathbf{z}} \to \hat{\mathbf{p}})$, see e.g., [100], carries each partial wave from an original quantization along the z-axis (expressed by the Clebsch-Gordan coefficient) into a quantization axis along $\mathbf{p}$.

The solutions for negative-energy states $E < -1$ are derived from Eq. (6.26) to (6.31) by substituting $\mathbf{p} \to -\mathbf{p}$, $E \to -E$, and $Z \to -Z$.

In analogy to the bound-state case, one may also derive approximate relativistic continuum wavefunctions corresponding to Eqs. (2.47) and (2.48), accurate to the order αZ in the relativistic corrections. These functions, denoted as Sommerfeld-Maue [102] or Furry [103] wavefunctions, have been widely applied in the literature [104]. As stated above, the nonrelativistic wave equation for continuum states in a Coulomb field is separable in parabolic coordinates, but the corresponding Dirac equation is not. Nevertheless, the Sommerfeld-Maue wavefunction, similarly as Eq. (6.21) for bound states, is directly related to the solution of the nonrelativistic problem in parabolic coordinates. It is an advantage of the Sommerfeld-Maue approximation to render a decomposition of the continuum wavefunction into partial waves unnecessary. The derivation is analogous to, albeit more complicated than, the derivation of the Darwin wavefunction [3, 99, 89]. If the wavefunction is normalized such that for a vanishing potential $V \to 0$ it merges into the plane-wave solution $\varphi \to \exp(i\,\mathbf{k}\cdot\mathbf{r})\,u^{(\rho)}(\mathbf{k})$, where $u^{(\rho)}(\mathbf{k})$ is the four-component spinor for a free particle with energy E and wave vector $\mathbf{k}$, then the solution for *outgoing* spherical electron waves, to be used for *electron absorption*, can be written as

$$
\varphi_{E,k}^{(+)} \;=\; e^{\frac{\pi}{2}\eta}\,\Gamma(1-i\eta)\,e^{i\,\mathbf{k}\cdot\mathbf{r}}\left(1 - i\frac{\hbar c}{2E}\,\boldsymbol{\alpha}\cdot\nabla\right)
$$

$$
\times\; {}_1F_1\left[i\eta\,,1\,;i(kr - \mathbf{k}\cdot\mathbf{r})\right]u^{(\rho)}(\mathbf{k}) \qquad (6.32)
$$

and for *incoming* spherical waves, to be used for *electron emission*, as

$$\varphi_{E,k}^{(-)} \;=\; e^{\frac{\pi}{2}\eta}\,\Gamma(1+\mathrm{i}\eta)\,e^{\mathrm{i}\,\mathbf{k}\cdot\mathbf{r}}\left(1-\mathrm{i}\frac{\hbar c}{2E}\,\boldsymbol{\alpha}\cdot\nabla\right)$$
$$\times\ _1F_1\big[-\mathrm{i}\eta,1;-\mathrm{i}(kr+\mathbf{k}\cdot\mathbf{r})\big]\,u^{(\rho)}(\mathbf{k}). \quad (6.33)$$

Here, the Sommerfeld parameter is $\eta = \alpha Z E/(\hbar c k)$. We note that the equations (6.32) and (6.33) are the (approximate) relativistic extensions of the nonrelativistic equations (2.47) and (2.48), which are retrieved if the term with $\boldsymbol{\alpha}\cdot\nabla$ is dropped.

Again, the solutions for negative-energy states are derived from Eqs. (6.32) and (6.33) by substituting $\mathbf{k}\to-\mathbf{k}$, $E\to-E$, and $Z\to-Z$.

As has been shown by Bethe and Maximon [104], the Sommerfeld-Maue wavefunctions are obtained by replacing $s = (\kappa^2 - \alpha^2 Z^2)^{1/2}$ with $|\kappa|$ and hence are good approximations for angular momenta $l \gg Z/137$. They can be used whenever the lowest partial waves $l \simeq 1$ are not important. This is the case, regardless of the nuclear charge Z, if total electron and positron energies are large compared to $m_e c^2$.

Chapter 7

Relativistic ion-atom collisions: General theory

In Chapter 5 we have discussed the electromagnetic field produced by a point charge moving with relativistic velocity and in Chapter 6 the relativistic motion of a Dirac electron in a Coulomb potential. In the present Chapter, we merge these results by analyzing the electron motion in the combined potential of the space-fixed target and of the moving projectile nucleus.

In Sec. 7.1, we first consider the Dirac equation for two Coulomb potentials one of which is moving. After defining a criterion for a "fast" relativistic collision in Sec. 7.2, we write down, in a preliminary fashion, the perturbative transition amplitudes for the basic processes: excitation, ionization, charge transfer, and – as a new process – electron-positron pair production. Turning to moderately energetic collisions, we formulate the two-center coupled-channel method for the relativistic case in Sec. 7.3. After finding the asymptotic solutions of the two-center Dirac equation in Sec. 7.4, we argue in Sec. 7.5 that basis states satisfying Coulomb boundary conditions should lead to a faster convergence, to be shown in subsequent Chapters. Finally, as applications of purely numerical methods, we refer to solutions on a lattice in position space in Sec. 7.6 and to solutions on a lattice in momentum space in Sec. 7.7.

7.1 Dirac equation for moving Coulomb potentials

In analogy to Eqs. (1.18) and (1.19), we write down the time-dependent Dirac equation for an electron moving in electromagnetic potentials generated by two Coulomb sources which pass each other at a relativistic speed.

This equation is the basis for the entire subsequent treatment. Taking the target nucleus to be at rest in the laboratory frame, we have

$$i\hbar \frac{\partial}{\partial t} \Psi(\mathbf{r}_{\mathrm{T}}, t) = H \Psi(\mathbf{r}_{\mathrm{T}}, t) \tag{7.1}$$

with

$$H = -i\hbar c \,\boldsymbol{\alpha} \cdot \nabla_{\mathrm{T}} - \frac{e^2 Z_{\mathrm{T}}}{r_{\mathrm{T}}} - S^2 \frac{e^2 Z_{\mathrm{P}}}{r'_{\mathrm{P}}(t)} + m_e c^2 \gamma^0. \tag{7.2}$$

Here, the 4×4 matrix S^2, given by Eq. (6.11), Lorentz-transforms the projectile Coulomb potential $-e^2 Z_{\mathrm{P}}/r'_{\mathrm{P}}(t)$ from the projectile frame into the target system in which the Dirac equation (7.1) is written [3]. The Hamiltonian (7.2) describes an electron subject to the stationary Coulomb potential $-e^2 Z_{\mathrm{T}}/r_{\mathrm{T}}$ of the target nucleus and to the time-dependent projectile potential

$$-S^2 \frac{e^2 Z_{\mathrm{P}}}{r'_{\mathrm{P}}(t)} = -\gamma (1 - \beta\alpha_z) \frac{e^2 Z_{\mathrm{P}}}{r'_{\mathrm{P}}(t)}. \tag{7.3}$$

This expression has a simple physical interpretation. In classical electrodynamics, the Lorentz-invariant interaction between a four-current $j^\mu = (c\rho, \mathbf{j})$ and a four-potential $A^\mu = (\Phi, \mathbf{A})$ is the scalar product

$$j_\mu A^\mu / c = \rho \Phi - \mathbf{j} \cdot \mathbf{A}/c. \tag{7.4}$$

Observing that

$$\rho = -e\,\psi^\dagger \psi \qquad \text{and} \qquad \mathbf{j} = -ec\,\psi^\dagger \,\boldsymbol{\alpha}\,\psi \tag{7.5}$$

are the Dirac expressions for the electron charge density and current density, respectively, and taking into account the Liénard-Wiechert potentials (5.32), we directly obtain the expression (7.3). A similar equation can be derived in the projectile system [3]. In this case, the target potential has to be transformed into the projectile frame with the operator $S^{-2}(v) = S^2(-v)$, see Eq. (6.11).

The potentials appearing in the two-center time-dependent Dirac equations (7.1) and (7.2) are of long range, so that conventional scattering theory is no longer applicable without special precautions. This has been discussed in Sec. 4.3. In general, it is convenient and sometimes necessary to eliminate from the initial- and final-state electronic wavefunctions the asymptotic time dependence caused by the long-range Coulomb interaction. Before getting to this problem in Sec. 7.4, let us confine ourselves, in Secs. 7.2 and 7.3, to the conventional treatment which is still widely applied.

7.2 Perturbative transition amplitudes

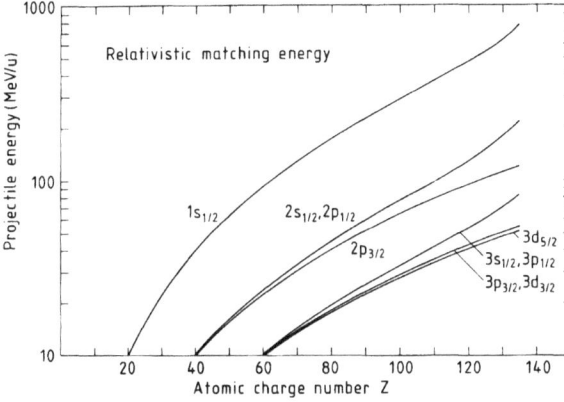

Figure 7.1: Collision energy in MeV/u for which energy matching according to Eq. (7.6) is obtained as a function of the atomic charge number for various electronic eigenstates. From [3].

For nonrelativistic collisions, it is convenient to distinguish low-energy and high-energy collisions, depending on whether the projectile velocity is smaller or larger than the velocity of the active electron, see Sec. 1.1 and Eq. (2.1). In the former case and in the transition region, the electron is able to follow the nuclear motion. For relativistic collisions, all velocities approach the speed of light, so that a comparison of velocities is no longer meaningful. In this case, it is more appropriate to define a *matching energy* T_m per atomic mass unit or the corresponding Lorentz factor $\gamma_m = 1 + T_m$ (MeV/u)/931.5, see Eq. (5.22), by demanding that the kinetic energy of a free electron travelling with the speed of the projectile, equals the binding energy E_{bind} of the electronic state, that is

$$m_e c^2 (\gamma_m - 1) = E_{bind} = m_e c^2 [1 - \sqrt{1 - (\alpha Z)^2}], \qquad (7.6)$$

where the second equation refers to atomic $1s_{1/2}$ states. The definition (7.6) has the correct nonrelativistic limit (2.1). Figure 7.1 shows the matching energy T_m calculated from Eq. (7.6) for different atomic states as a function of the charge number Z. There are other possible prescriptions leading to similar estimates.

For example, for a collision involving $1s_{1/2}$ states of uranium, the matching energy of the projectile is $T_m = 240$ MeV/u, so that "fast" collisions start from, say, 1 GeV/u. Above this energy, we expect perturbative methods to be applicable. For fast collisions, we may apply the Born expansion of the

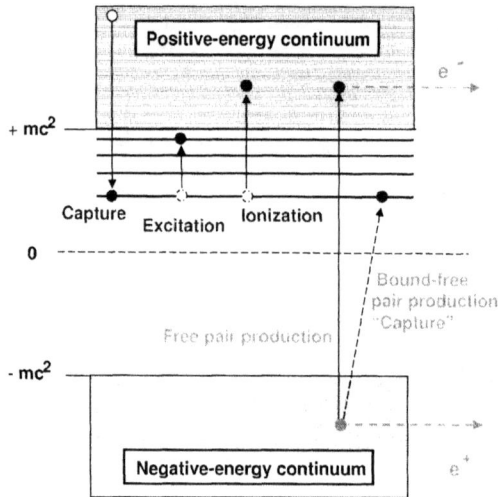

Figure 7.2: Schematic illustration of the main atomic processes occurring in relativistic ion-atom collisions. From [105]

transition amplitude as has been dicussed in Sec. 3.1. The only difference for relativistic collisions is that the interaction is now given by Eq. (7.4) as $j_\mu A^\mu/c$ or in the Dirac form by Eq. (7.3).

In Fig. 7.2 we schematically indicate the main atomic processes that can occur in a relativistic ion-atom collision. It is noted that, in contrast to the nonrelativistic case, the Dirac spectrum exhibits a negative-energy continuum in addition to the positive-energy continuum. However, except for pair production, the negative-energy continuum plays no direct role. Capture – actually from the bound state of another atom – excitation and ionization are governed by the upper half of the diagram, similarly as in the nonrelativistic case.

In the following, we list the first-order relativistic transition amplitudes of these processes applicable to energetic collisions with energies considerably higher than the matching energy T_m of Fig. 7.1.

(a) *Excitation and ionization*
Since in the projectile system, the interaction is simply Coulombic, the
perturbation by the projectile, Eq. (7.3), directly leads to the result

$$A_{fi}^{e} = -i\frac{e^2}{\hbar} \int dt \int d^3r_{\mathrm{T}}\, \psi_f^{\dagger}(\mathbf{r}_{\mathrm{T}},t)\, S^2 \left(-\frac{Z_{\mathrm{P}}}{r_{\mathrm{P}}'}\right) \psi_i(\mathbf{r}_{\mathrm{T}},t), \qquad (7.7)$$

which corresponds to the nonrelativistic amplitude (3.25).
(b) *Charge transfer*
Because the final-state electronic wavefunction is transformed from the
projectile frame to the target system with S^{-1}, the transformation matrix
S occurs only in first power, so that

$$A_{fi}^{tr} = -i\frac{e^2}{\hbar} \int dt \int d^3r_{\mathrm{T}}\, \psi_f'^{\dagger}(\mathbf{r}_{\mathrm{P}}',t')\, S \left(-\frac{Z_{\mathrm{P}}}{r_{\mathrm{P}}'}\right) \psi_i(\mathbf{r}_{\mathrm{T}},t). \qquad (7.8)$$

This amplitude corresponds to the nonrelativistic version (4.2). Both in
Eqs. (7.8) and (4.2), Coulomb boundary conditions are not yet included.
This will be done in Sec. 7.3.
(c) *Electron-positron pair creation*
The transition amplitude is similar as for excitation, namely

$$A_{fi}^{pair} = -i\frac{e^2}{\hbar} \int dt \int d^3r_{\mathrm{T}}\, \psi_{p_+}^{\dagger}(\mathbf{r}_{\mathrm{T}},t)\, S^2 \left(-\frac{Z_{\mathrm{P}}}{r_{\mathrm{P}}'}\right) \psi_{p_-}(\mathbf{r}_{\mathrm{T}},t), \qquad (7.9)$$

however, with the initial wavefunction replaced by the positron wavefunction ψ_{p_-} and the final wavefunction replaced by the electron wavefunction
ψ_{p_+}. The independent integration variables in all cases are $\mathbf{r}_{\mathrm{T}}$ and t, so that
the Lorentz-transformed coordinates (5.29) with (5.30) have to be inserted
for $\mathbf{r}_{\mathrm{P}}'$ and t' when the integrals are evaluated. The Lorentz transform
replaces the Galileo transform expressed by the translation factor (2.19)
for nonrelativistic collisions. Further approximations depend on the kind
of wavefunction actually inserted for the four-spinors ψ_i and ψ_f. From the
transition amplitudes, one may derive cross sections in the usual way, see
Sec. 3.4.

7.3 Two-center coupled-channel methods

For energies of the order of T_m or below, perturbation theory is no longer
applicable, hence one has to resort to nonperturbative methods. In analogy to nonrelativistic collisions, see Secs. 2.1 and 2.9, an efficient method
consists in solving the time-dependent Dirac equation (7.1) by adopting
a basis expansion. For cases like excitation and ionization as well as for

pair production, single-center expansions around the target nucleus have been applied. However, more generally, relativistic ion-atom collisions require the expansion of the time-dependent wavefunction $\Psi(\mathbf{r}_\mathrm{T}, t)$ in terms of target and projectile eigenstates $\psi_k(\mathbf{r}_\mathrm{T}, t)$ and $\psi_{k'}(\mathbf{r}'_\mathrm{P}, t')$, respectively, as

$$\Psi(\mathbf{r}_\mathrm{T}, t) = \sum_k a_k(t) \psi_k(\mathbf{r}_\mathrm{T}, t) + \sum_{k'} a_{k'}(t)\, S^{-1} \psi_{k'}(\mathbf{r}'_\mathrm{P}, t'), \qquad (7.10)$$

which replaces Eq. (2.38). Here, we have referred the total wavefunction to the laboratory system. The first sum represents an expansion in terms of target states; the second sum is an expansion in terms of projectile states originally defined in the projectile system and subsequently transformed to the laboratory system with the aid of the operator S^{-1} defined in Eqs. (6.8) and (6.9).

The use of two centers for the expansion of the wavefunction $\Psi(\mathbf{r}, t)$ has two important implications: (a) In the interaction region, linear combinations of atomic orbitals around the target and around the projectile nucleus provide flexibility to approximate transient molecular orbitals which may play a role during the collision. (b) Charge transfer is explicitly included as a reaction channel. This not only renders charge exchange at all possible, but will also affect other reactions by entering via a set of intermediate states.

The only approximation within the method consists in the restriction to a finite set of atomic basis states. Since each of the atomic basis sets is truncated, there is no problem with overcompleteness that would arise from two complete (or nearly complete) sets. In practical calculations [106, 107], basis sets have been confined to 20 or 36 substates, namely those of the $1s_{1/2}$, $2s_{1/2}$, $2p_{1/2}$, $2p_{3/2}$, $3s_{1/2}$, $3p_{1/2}$, and $3p_{3/2}$ shells at each center, see Fig. 7.2 and Sec. 9.3. For a general discussion of basis states, see Sec. 2.1.

In Eq. (7.10) the Lorentz transformation of the projectile time factors $\exp(-\mathrm{i}\, E_{k'} t')$, where t' transforms according to Eq. (5.29), automatically implies the relativistic counterparts of the nonrelativistic translation factors, see Eq. (2.19).

In addition to bound target and projectile eigenstates $\psi_k(\mathbf{r}_\mathrm{T}, t)$ and $\psi_{k'}(\mathbf{r}'_\mathrm{P}, t')$, respectively, one may also include in Eq. (7.10) discretized continuum states constructed as superpositions of adjacent continuous eigenstates $\varphi_E(\mathbf{r})$, see e.g., [3]. Stationary "wave packets" (i.e., built from *time-independent* eigenstates) centered around E_k and with the width ΔE_k are represented by $\psi_{E_k}(\mathbf{r}, t) = \varphi_{E_k}(\mathbf{r}) \exp(-\mathrm{i}E_k t/\hbar)$, where

$$\varphi_{E_k}(\mathbf{r}) = \frac{1}{\sqrt{\Delta E_k}} \int_{E_k - \Delta E_k/2}^{E_k + \Delta E_k/2} \varphi_E(\mathbf{r})\, \mathrm{d}E. \qquad (7.11)$$

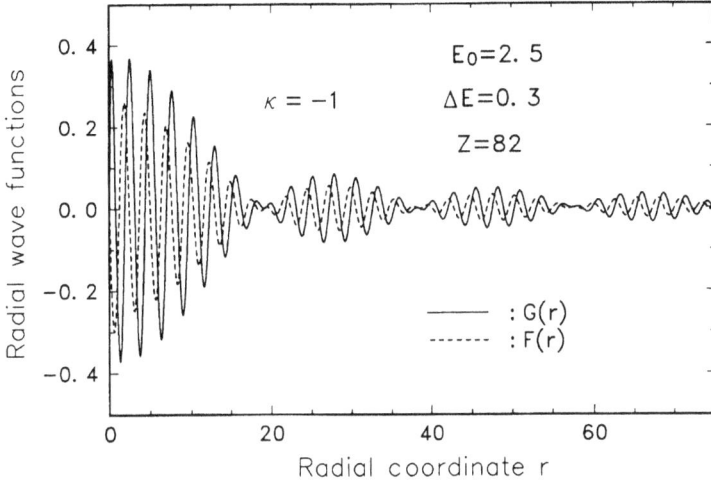

Figure 7.3: The radial components $G(r) = r\,g(r)$ and $F(r) = r\,f(r)$ of relativistic wave packets (in relativistic units) are plotted versus the radial coordinate r for the $s_{1/2}$-state ($\kappa = -1$). The charge is taken as $Z = 82$, the energy in Eq. (7.11) as $E_k = 2.5$ and the width of the integration interval as $\Delta E_k = 0.3$. From [3].

These wavepackets are orthogonal to bound eigenstates φ_{κ,m_j} and among each other, because of the orthogonality property of eigenstates, see Eq. (6.27) for the continuum states $\varphi_{E_k}(\mathbf{r})$, and because the intervals in Eq. (7.11) do not overlap.

The coordinate $\mathbf{r}$ in Eq. (7.11) has to be identified with the target or the projectile coordinate $\mathbf{r}_{\mathrm{T}}$ or $\mathbf{r}'_{\mathrm{p}}$, respectively. Owing to the superposition of adjacent continuum functions, the fall-off of the radial wave functions of (7.11) is much faster than the decrease $1/r$ of the radial eigenfunctions $g(r)$ and $f(r)$ in Eq. (6.28) themselves, see Fig. 7.3. The inclusion of discretized continuum functions is of importance not only for the description of ionization and pair production, but also will affect bound-state reactions like charge transfer.

Proceeding as in the nonrelativistic problem, Sec. 2.9, we may write the set of coupled equations, in analogy to Eq. (2.42), in the matrix form

$$\dot{\mathbf{a}} = -\frac{i}{\hbar}\,\mathbf{N}^{-1}\,\mathbb{V}\,\mathbf{a}. \tag{7.12}$$

As in Eq. (2.42), the overlap matrix $\mathbb{N}$ and the interaction matrix $\mathbb{V}$ are each built from four submatrices corresponding to target-target, target-projectile, projectile-target, and projectile-projectile transitions [106, 107].

The overlap matrices, replacing the expressions (2.40), are

$$
\begin{aligned}
N_{ik} &= \delta_{ik} \\
N_{ik'} &= \int \varphi_i^\dagger(\mathbf{r_T}) S^{-1} \varphi_{k'}(\mathbf{r_P'}) \, e^{i\eta E_{k'}(v/\hbar c^2)z_\mathrm{T}} \; d^3 r_\mathrm{T} \; e^{-i(\gamma E_{k'} - E_i)t/\hbar} \\
N_{i'k} &= \int \varphi_{i'}^\dagger(\mathbf{r_P'}) S^{-1} \varphi_k(\mathbf{r_T}) \, e^{-i\eta E_{i'}(v/\hbar c^2)z_\mathrm{T}} \; d^3 r_\mathrm{T} \; e^{-i(E_k - \gamma E_{i'})t/\hbar} \\
N_{i'k'} &= \delta_{i'k'}.
\end{aligned}
\tag{7.13}
$$

The single-center overlap matrix elements (target-target and projectile-projectile) are time-independent while the two-center matrix elements (target-projectile and projectile-target) are time-dependent owing to the relative motion of the collision partners.

In analogy to the overlap, we have four types of interaction matrix elements, replacing the expressions (2.41), namely

$$
\begin{aligned}
V_{ik} &= \int \varphi_i^\dagger(\mathbf{r_T}) \left(-S^2 \frac{e^2 Z_\mathrm{P}}{r_\mathrm{P}'} \right) \varphi_k(\mathbf{r_T}) \, d^3 r_\mathrm{T} \; e^{-i(E_k - E_i)t/\hbar} \\
V_{ik'} &= \int \varphi_i^\dagger(\mathbf{r_T}) \left(-S^{-1} \frac{e^2 Z_\mathrm{T}}{r_\mathrm{T}} \right) \varphi_{k'}(\mathbf{r_P'}) \\
&\qquad \times \, e^{i\eta E_{k'}(v/\hbar c^2)z_\mathrm{T}} \; d^3 r_\mathrm{T} \; e^{-i(\gamma E_{k'} - E_i)t/\hbar} \\
V_{i'k} &= \int \varphi_{i'}^\dagger(\mathbf{r_P'}) \left(-S \frac{e^2 Z_\mathrm{P}}{r_\mathrm{P}'} \right) \varphi_k(\mathbf{r_T}) \\
&\qquad \times \, e^{-i\eta E_{i'}(v/\hbar c^2)z_\mathrm{T}} \; d^3 r_\mathrm{T} \; e^{-i(E_k - \gamma E_{i'})t/\hbar} \\
V_{i'k'} &= \int \varphi_{i'}^\dagger(\mathbf{r_P'}) \left(-S^{-2} \frac{e^2 Z_\mathrm{T}}{r_\mathrm{T}} \right) \varphi_{k'}(\mathbf{r_P'}) \\
&\qquad \times \, e^{-i\eta(E_{k'} - E_{i'})(v/\hbar c^2)z_\mathrm{T}} \; d^3 r_\mathrm{T} \; e^{-i\eta(E_{k'} - E_{i'})t/\hbar}.
\end{aligned}
\tag{7.14}
$$

One verifies that in the nonrelativistic limit, all matrix elements merge into the corresponding nonrelativistic expressions including translation factors. The initial conditions are given by Eq. (2.43) and the resulting cross sections in analogy to Eqs. (3.23) and (3.24). For charge transfer, the label k has to be replaced by the label k'.

For the collision of U^{92+} ions with U^{91+} targets at 0.5 and 1.0 GeV/u, experimental data are not available but, on the other hand, the collision is well suited to serve as a model case for which effects of relativistic electron and relativistic projectile motion can be studied. The energy chosen is well above the matching energy given in Fig. 7.1, but this is still within the proper range for applying the coupled-channels method.

Figure 7.4 shows some results for the time evolution of the occupation

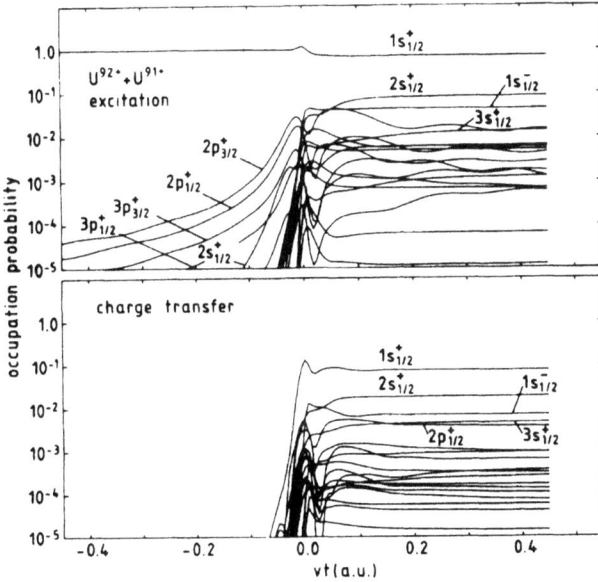

Figure 7.4: Time evolution of the occupation probabilities of target states in excitation and of projectile states in charge transfer in $U^{92+} + U^{91+}$ collisions at 500 MeV/u laboratory energy and impact parameter $b = 0.01$ a.u. $\approx a_{\rm K}$. The abscissa plotted is the projection of the projectile-target separation on the beam direction. The projection of the angular momentum is indicated by $+,-$, and $++$ for $\mu_j = \frac{1}{2}, -\frac{1}{2}$, and $\frac{3}{2}$, respectively. From [107].

probabilities for target states (excitation) and projectile states (charge transfer) for the impact energy of 500 MeV/u using 36 atomic basis states up to the $3p_{3/2}$ shell of target and projectile and assuming an impact parameter of 0.01 a.u. This distance is approximately equal to the K-shell radius $a_{\rm K}$ and roughly corresponds to the region of maximum contribution to the excitation cross section. The range $|vt| \leq 40\, a_{\rm K}$, displayed in Fig. 7.4, represents a narrow window out of the total time range $|vt| \leq (500 - 2000)\, a_{\rm K}$ actually treated in the calculation [107].

The figures reveal certain qualitative features: (a) The importance of multistep processes is shown by the rapid fluctuations of some of the occupation probabilities. (b) Electron transfer, both during the collision as well as in the exit channel, is comparable to excitation and cannot be neglected. (c) With increasing shell size, the interaction region extends much farther out. (d) In this treatment, certain target states are affected by the projectile long before and long after it reaches its distance of closest approach at $t = 0$. This is clearly caused by the long range of the Coulomb interaction. The elimination of these couplings is discussed in Secs. 8.2 and 8.3.

7.4   Asymptotic solutions

As has been indicated in Sec. 4.3 for the nonrelativistic case and in Sec. 7.1 for relativistic collisions, the long range of the Coulomb interaction and consequently of the Liénard-Wiechert interaction requires a special treatment in order to ensure that perturbation theory is applicable. This is done by imposing appropriate Coulomb boundary conditions, which amounts to applying an appropriate gauge transformation, so that the resulting transition amplitude contains *short range* couplings. Here "short range" means that the perturbation falls off more rapidly than $1/r$ as a function of the electron-nucleus separation.

In order to establish well-defined channel wavefunctions that are devoid of the asymptotic time dependence caused by long-range interactions, we have to derive the asymptotic solutions of Eqs. (7.1) and (7.2) for $t \to \pm\infty$. Referring to Fig. 4.2 introduced for nonrelativistic collisions, we notice that if the electron is bound to one of the collision partners, its asymptotic separation from the other partner almost equals the internuclear separation, so that

$$\lim_{t \to -\infty} r'_{\rm P} \;=\; R' = \sqrt{b^2 + v^2 t'^2}$$
$$\lim_{t \to +\infty} r_{\rm T} \;=\; R = \sqrt{b^2 + v^2 t^2}. \tag{7.15}$$

Here, t and $t' = \gamma(t - vz_{\rm T}/c^2)$ are the electronic time coordinates, referred to the target and projectile frame, respectively. Thus R and R' measure the internuclear separations at times associated with the electronic positions $\mathbf{r}_{\rm T}$ or $\mathbf{r}'_{\rm P}$, respectively. In Fig. 4.2, the primes at the coordinates $\mathbf{r}'_{\rm p}$ and $\mathbf{R}'$ refer to the projectile Lorentz system.

In the limits expressed by Eqs. (7.15), the following asymptotic equations in the target frame have to be satisfied [3]: For an electron asymptotically bound to the *target* nucleus at $t \to -\infty$, we have, in analogy to Eqs. (4.14),

$$i\hbar \frac{\partial}{\partial t} \Phi_T^\infty(\mathbf{r}_T, t) = \left(-i\hbar c\boldsymbol{\alpha} \cdot \nabla_T - \frac{Z_T e^2}{r_T} - S^2 \frac{Z_P e^2}{R'} + m_e c^2 \gamma^0 \right)$$

$$\times \Phi_T^\infty(\mathbf{r}_T, t), \qquad (7.16)$$

and for an electron asymptotically bound to the *projectile* nucleus at $t \to +\infty$,

$$i\hbar \frac{\partial}{\partial t} \Phi_P^\infty(\mathbf{r}_T, t) = \left(-i\hbar c\boldsymbol{\alpha} \cdot \nabla_T - \frac{Z_T e^2}{R} - S^2 \frac{Z_P e^2}{r_P'} + m_e c^2 \gamma^0 \right)$$

$$\times \Phi_P^\infty(\mathbf{r}_T, t). \qquad (7.17)$$

Similar equations can be written down in the projectile system [3].

In order to solve these equations, we introduce the dimensionless Sommerfeld parameters

$$\nu_T = \frac{e^2 Z_T}{v\hbar} \quad \text{and} \quad \nu_P = \frac{e^2 Z_P}{v\hbar}. \qquad (7.18)$$

and the time-dependent unperturbed eigenfunctions subject to the single-center Dirac equations in the target or in the projectile system, respectively,

$$i\hbar \frac{\partial}{\partial t} \psi_T(\mathbf{r}_T, t) = \left(-i\hbar c\boldsymbol{\alpha} \cdot \nabla_T - \frac{Z_T e^2}{r_T} + m_e c^2 \gamma^0 \right) \psi_T(\mathbf{r}_T, t) \qquad (7.19)$$

and

$$i\hbar \frac{\partial}{\partial t'} \psi_P'(\mathbf{r}_P', t') = \left(-i\hbar c\boldsymbol{\alpha} \cdot \nabla_P' - \frac{Z_P e^2}{r_P'} + m_e c^2 \gamma^0 \right) \psi_P'(\mathbf{r}_P', t'). \qquad (7.20)$$

With these definitions, it follows that the exact asymptotic solutions of Eqs. (7.16) and (7.17), respectively, are obtained by a phase transformation in a factorized form. The resulting solutions, first discussed in the relativistic case by Eichler [108], satisfy Coulomb boundary conditions (more precisely: Liénard-Wiechert boundary conditions), because they take into account the asymptotic long-range potentials caused by the distant charges and can be written[†] as

---

[†]Various existing forms of the phase are equivalent, if they differ only by an additive constant phase. Note, for example, that $b^2 = (R + vt)(R - vt)$ and hence $\ln(R + vt) = -\ln(R - vt) + \text{const.}$

$$\Phi_T^\infty(\mathbf{r}_T,t) = e^{-i\nu_P \ln(vR'-v^2t')}\psi_T(\mathbf{r}_T,t) \tag{7.21}$$

and

$$\Phi_P^\infty(\mathbf{r}_T,t) = e^{i\nu_T \ln(vR+v^2t)}S^{-1}\psi_P'(\mathbf{r}_P',t'), \tag{7.22}$$

corresponding to the nonrelativistic version (4.16).

The phase factor of Eq. (7.21), in contrast to its counterpart in Eq. (7.22), carries a dependence on the space coordinates in the laboratory system through the relation $t' = \gamma(t - vz_T/c^2)$. In generalization of Sec. 4.3, the phase transformations eliminate the asymptotic limits of the scalar and vector potentials (5.32). This may also be regarded as a gauge transformation.

7.5   Basis states satisfying Coulomb boundary conditions

In coupled-channel calculations, the choice of the basis states is a matter of convenience, both for the desired interpretation of the results and for the ease of calculations. In Eq. (7.10), we have chosen the conventional view that the basis states are stationary atomic orbitals. In the light of the discussion in Sec. 7.4, it appears more appropriate to use a set of basis states that satisfy Coulomb boundary conditions. We therefore replace Eq. (7.10) with the expansion [109]

$$\Psi(\mathbf{r}_T,t) = \sum_k a_k(t)e^{-i\nu_P \ln(vR'-v^2t')}\psi_k(\mathbf{r}_T,t)$$

$$+ \sum_{k'} a_{k'}(t)e^{i\nu_T \ln(vR+v^2t)}S^{-1}\psi_{k'}(\mathbf{r}_P',t'). \tag{7.23}$$

The first term represents an expansion in terms of target states asymptotically phase-distorted by the projectile, while the second sum is an expansion in terms of projectile states asymptotically phase-distorted by the target.

When inserting this expansion into Eq. (7.1), we find that on the right-hand side, the time differentiation of the logarithmic phase factors yields additional terms so that we get the replacement

$$\frac{e^2 Z_P}{r_P'} \rightarrow \frac{e^2 Z_P}{r_P'} - \frac{e^2 Z_P}{R'}$$

$$\frac{e^2 Z_T}{r_T} \rightarrow \frac{e^2 Z_T}{r_T} - \frac{e^2 Z_T}{R} \tag{7.24}$$

leading to short-range interactions. The phase factors themselves drop out in the one-center integrals of Eqs. (7.13) and (7.14) since they are state-independent, but they remain in the two-center integrals. With these substitutions it is obvious how to modify the overlap matrix elements (7.13) and the potential matrix elements (7.14). It becomes apparent that with the basis expansion of (7.23), the long range of the Coulomb interaction is explicitly taken into account and hence the long-range couplings disappear. It is shown in Secs. 7.6, 8.2, and in 9.5 that the use of a phase-distorted basis set speeds up the convergence.

7.6 Numerical solutions on a lattice in position space

Perturbative calculations are known to grossly overestimate the probability for ionization and excitation at intermediate relativistic velocities and small impact parameters, when the perturbing potential becomes very strong during the close passage. One remedy is the nonperturbative treatment by coupled-channel calculations, see Secs. 7.3 and 7.5. Another nonperturbative method consists in the direct numerical solution of the time-dependent Dirac equation (7.1) on a lattice.

We first consider the finite-difference method in position space and subsequently proceed to momentum space. Adopting this purely numerical method [110, 111, 112], the Dirac equation (7.1) is solved on a lattice departing from Eq. (7.12) which, for a small time interval Δt, can be written as

$$\mathbf{a}(t + \Delta t) = \exp\left[-i\mathbb{N}^{-1}\mathbb{V}\Delta t/\hbar\right]\mathbf{a}(t), \qquad (7.25)$$

where $\mathbb{V}$ is assumed to be constant during the time interval Δt. The exponential operator is expanded in a power series in Δt up to the order n according to

$$\exp(-x) = \left(1 - x\left(1 - \frac{x}{2}\left(1 - \frac{x}{3}\left(\cdots - \frac{x}{n}\right)\cdots\right)\right)\right). \qquad (7.26)$$

With this method, a high numerical accuracy has been achieved [112].

In [112], the colliding system, carrying an electron initially bound in the 1s-state of the target, has been enclosed in a cartesian box with a side length of about 20 K-shell radii a_K for uranium. In order to achieve a rapid decrease of the transition probabilities for large internuclear separations, the long-range potentials have been reduced to short range by the phase transformations (7.21) and (7.22). Compared to earlier calculations [110, 111], the phase transformation greatly improves the convergence. Furthermore, symmetries have been exploited to reduce the numerical effort.

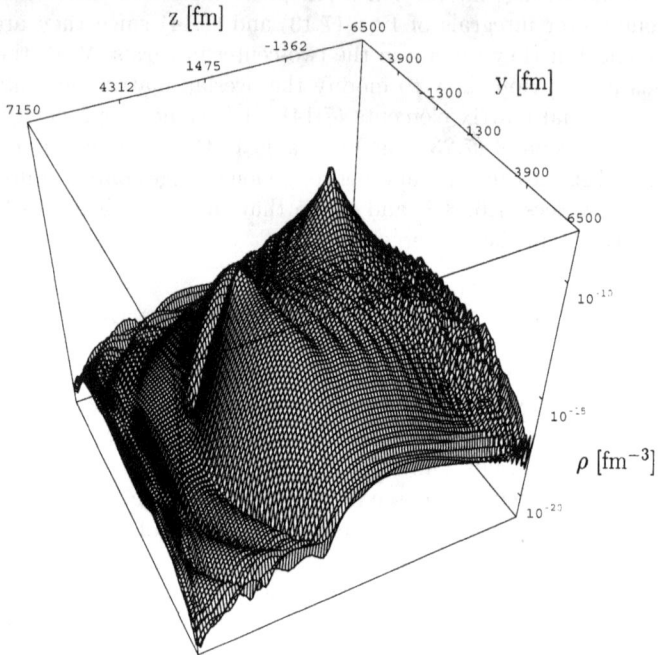

Figure 7.5: Probability density for the time-developed electron state $1s_{1/2}(m_j = 1/2)$ in the scattering plane for the collision of $U^{92+}(\gamma = 1.5 \equiv 466$ MeV/u) on U^{91+} at the time $t = 4780$ fm/$c \approx 8.3$ a_K/c at an impact parameter $b = 1060$ fm ≈ 1.8 a_K. The front left peak is the transferred density around the projectile nucleus, the back right peak is the remaining density around the target nucleus. From [112].

More than 2600 time steps have been taken into account, and the evolution operator in (7.25) has been expanded up to to $n = 7$. The electron density distribution at the end of the time evolution is displayed in Fig. 7.5. It shows the transferred density around the projectile together with the remaining density around the target nucleus.

In order to interpret the final density amplitude, it is necessary to numerically project on bound target states, in this case with principal quantum numbers $n \leq 7$ in order to obtain the exitation probabilities. For the ionization probabilities, projections on states of the positive energy continuum have been performed up to $|\kappa| \leq 5$ and up to an energy of 4.3

$m_e c^2$. Within this approach, also contributions of negative-energy states have been calculated. This will be discussed in Sec. 11.2.

7.7 Numerical solutions on a lattice in momentum space

A technical problem for the grid calculations in position space is the oscillatory behavior of the non-stationary non-localized wavefunctions. While these wavefunctions extend over the whole space, one sets them zero at the lattice boundary in practical calculations [111]. In order to avoid this problem, one may integrate the time-dependent Dirac equation on a lattice in momentum space [113, 114]. An advantage of this method, compared to the solution in position space, consists in the localization of the momentum wavefunction around the origin during the whole collision. Indeed, for the limiting case of a free wave, the momentum representation is a delta function, hence, for a distorted continuum wave, the momentum spread will still be small. As a result, the continuum wavefunction almost vanishes at the boundary of the box (in momentum space) enclosing the grid, so that there are no problems with boundary conditions.

Assuming a target nucleus at rest, a rectilinear trajectory $\mathbf{R} = (b, 0, vt)$, and adopting relativistic units, one obtains the time-dependent space wavefunction $\psi(\mathbf{k}, t)$ from the momentum wavefunction as [115]

$$\psi(\mathbf{r}, t) = \frac{1}{(2\pi)^3} \int d^3 k \, \psi(\mathbf{k}, t) \, e^{i\mathbf{k}\cdot\mathbf{r}}. \tag{7.27}$$

Insertion of this definition into the two-center Dirac equation (7.2), yields an integro-differential equation in momentum space

$$
i\frac{\partial}{\partial t}\psi(\mathbf{k}, t) = (\boldsymbol{\alpha} \cdot \mathbf{k} + \gamma^0)\psi(\mathbf{k}, t)
$$
$$
- \frac{\alpha}{2\pi} \int d^3 k' \left[\frac{Z_{\mathrm{T}}}{q^2} + \frac{Z_{\mathrm{P}}(1 - \beta\alpha_z)\, e^{i\mathbf{q}\cdot\mathbf{R}(t)}}{q_x^2 + q_y^2 + q_z^2/\gamma^2} \right] \psi(\mathbf{k}, t), \tag{7.28}
$$

with the initial condition

$$\lim_{t \to -\infty} = \phi_j(\mathbf{k})\, e^{-iE_j t}, \tag{7.29}$$

where the initial-state wavefunction ϕ_j describes a target K-shell electron with energy E_j. The vector $\mathbf{q} = \mathbf{k} - \mathbf{k}'$ represents the momentum transfer. The two terms in the square bracket in Eq. (7.28) are the Fourier transforms of the target and projectile potentials, respectively.

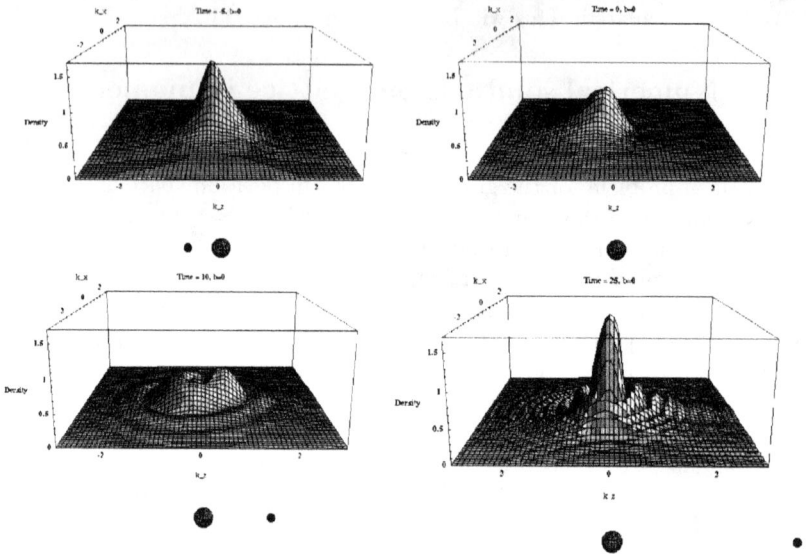

Figure 7.6: Electron density in momentum space at various time steps during a collision $Au^{79+} + U^{91+}$ with an energy of 0.96 GeV/u. The position of the target nucleus during the collision remains unchanged at $k = 0$. The projectile, indicated by the small sphere in coordinate space, moves along the k_z-axis from the negative to the positive direction. From [105].

The calculation of the momentum integral requires a very fine mesh and hence extensive computing time. As an example for these calculations, Fig. 7.6 shows the scalar momentum density $|\psi(\mathbf{k}, t)|^2$ in the reaction plane during a $Au^{79+} + U^{91+}$ collision at 0.96 GeV/u and for an impact parameter of $b = 0$. At the beginning of the collision, $t = -5$ (upper left picture), the electron is bound in a $1s_{1/2}$-state of the uranium ion. During the approach of the nuclei, there is almost no change of the electron momentum density, in contrast to the space representation of Sec. 7.6. Even at the time of closest approach, $t = 0$ (upper right picture), not too much has changed in the momentum distribution. Because of the attraction by the projectile, the momentum distribution is shifted in the negative k_z-direction. When the projectile has passed the target ion, $t = 10$, (lower left picture) transitions into states with higher momenta occur, so that the vicinity of the origin is emptied. Later on, at $t = 25$ (lower right picture), part of the density flows back to the origin. The probability for the various processes as excitation, ionization, charge transfer, and pair production can be obtained by numerically projecting the numerical wavefunction on momentum eigenstates of target or projectile, similarly as for the treatment in position space.

The grid calculations presented in Secs. 7.6 and 7.7, while yielding a vivid picture of the collision, are numerically very demanding and hence have not been widely applied.

Chapter 8

Direct reactions: Excitation and ionization

In this Chapter, we apply the general methods outlined in Chapter 7 to direct reactions, that is to excitation and ionization of the target atom by a projectile ion or vice versa. We start in Sec. 8.1 with a description in the framework of perturbation theory and, in particular, refer to the contributions of magnetic transitions absent in nonrelativistic collisions. For projectile excitation, we briefly discuss the influence of target electrons. The effects of long-range couplings in perturbation theory are investigated in Sec. 8.2 along with a way to eliminate them. In Sec. 8.3, we examine two-center coupled-channel calculations with and without imposing Coulomb boundary conditions. We also refer to calculations on a lattice.

8.1 First-order perturbation theory

For a fast collision and for impact parameters b comparable with, or larger than, the atomic K-shell radius, the transient perturbation on a K-shell electron of the target atom by the projectile is small enough that the first-order time-dependent perturbation theory is expected to be a good approximation, even for a high-Z projectile.

The transition amplitude (7.7) replacing the nonrelativistic expression (3.25) for excitation and ionization of the target electron is written as

$$A_{\mathrm{fi}}^{\mathrm{e}}(b) = \mathrm{i}\frac{\gamma e^2 Z_{\mathrm{P}}}{\hbar} \int \mathrm{d}t \, \mathrm{e}^{\frac{\mathrm{i}}{\hbar}(E_{\mathrm{f}} - E_{\mathrm{i}})t} \int \mathrm{d}^3 r_{\mathrm{T}} \, \varphi_{\mathrm{f}}^{\dagger}(\mathbf{r}_{\mathrm{T}})(1 - \beta\alpha_z)\frac{1}{r_{\mathrm{P}}'} \varphi_{\mathrm{i}}(\mathbf{r}_{\mathrm{T}}).$$

$$(8.1)$$

Here, φ_i and φ_f are the initial and final eigenfunctions of the target electron with energies E_i and E_f, while r'_P, given by Eq. (5.30) [where the unprimed coordinates are defined with respect to the target], is the electron-projectile distance measured in the projectile rest system. The cross sections for excitation and ionization resulting from the transition amplitudes are given by Eqs. (3.23) and (3.24), respectively.

For the evaluation of the amplitude (8.1), it is convenient to rewrite the retarded interaction in terms of its Fourier transform. We introduce the relativistically generalized Bethe integral [3, 116, 117]

$$\frac{1}{r'_P} = \frac{1}{2\pi^2\gamma} \int \frac{e^{i\mathbf{q}\cdot(\mathbf{r}_T - \mathbf{R})}}{q^2 - \beta^2 q_z^2}\, d^3q \qquad (8.2)$$

and the abbreviation

$$q_0 = \frac{E_f - E_i}{\hbar v}, \qquad (8.3)$$

which is connected with the nonrelativistic minimum momentum transfer $\hbar q_0$. Furthermore, decomposing $\mathbf{q}$ as $\mathbf{q} = (\mathbf{q}_b, q_z)$, where $\mathbf{q}_b = \mathbf{q}_\perp$ is the component perpendicular to the beam direction $\hat{\mathbf{z}}$, and performing the time integration in Eq. (8.1), one obtains the transition amplitude

$$A_{fi}^e(b) = \frac{i}{\pi}v_P \int \frac{d^2q_b}{q_b^2 + (1 - \beta^2)q_0^2}\, e^{-i\,\mathbf{q}_b \cdot \mathbf{b}}\, M_{fi}(\mathbf{q}) \qquad (8.4)$$

with

$$M_{fi}(\mathbf{q}) = \langle f \mid (1 - \beta\alpha_z)\, e^{i\,\mathbf{q}\cdot\mathbf{r}_T} \mid i\rangle. \qquad (8.5)$$

The advantage of this formulation is that the complicated time dependence contained in the expression (5.30) for r'_P is removed and replaced with the more explicit wave number dependence (8.4) and (8.5).

Ionization

The initial bound state is characterized by the quantum numbers $i \equiv \{n_i, \kappa_i, \mu_{j_i}\}$. For the final state with a given continuum energy E_f, one has to distinguish two possible situations determined by the experimental setup, namely whether the direction $\hat{\mathbf{k}}$ of emission of the electron is observed or is not observed. In the former case, one has a coherent, in the latter case an incoherent summation over partial waves of the continuum wavefunction in the final state.

The total energy-differential cross section for ionization with the electron direction undetected is written explicitly as

$$\frac{d\sigma_{fi}^e}{dE_f}\bigg|_{\text{total}} = 8\pi\nu_P^2 \frac{1}{2j_i + 1} \sum_{\mu_{j_i}} \sum_{\kappa_f} \sum_{\mu_{j_f}} \int_0^\infty \frac{q_b \, dq_b}{[q_b^2 + (1 - \beta^2)q_0^2]^2}$$

$$\times \left| \langle E_f, \kappa_f, \mu_{j_f} | (1 - \boldsymbol{\beta} \cdot \boldsymbol{\alpha}) \, e^{i \, \mathbf{q} \cdot \mathbf{r}_T} | n_i, \kappa_i, \mu_{j_i} \rangle \right|^2, \qquad (8.6)$$

where $\beta = \mathbf{v}/c$.

The experimental studies of projectile ionization and excitation in the relativistic velocity regime focus in general on collisions, in which high-Z one- and two-electron projectiles and low-Z targets are involved. Here, the perturbation caused by the nuclear charge of the target is small and, consequently, first-order perturbation theories should be appropriate with respect to the target. Also, the relativistic electron wavefunctions are exactly known for hydrogen-like ions and the assumption that the projectile moves on a straight-line trajectory is valid. Therefore, relativistic collisions are a clean testing ground for first-order perturbation theories, so that the effects caused by the magnetic part of the Liénard-Wiechert potential (5.32) can be examined in these experiments.

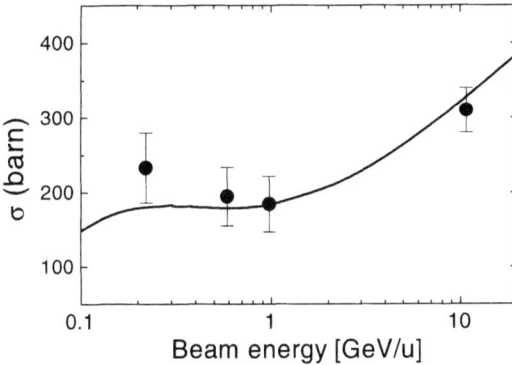

Figure 8.1: Cross sections for projectile ionization of H-like gold ions Au^{78+} colliding with carbon targets. From [118]

For ionization, total cross sections were studied systematically for a wide range of high-Z ions starting from xenon up to uranium. As an example, in Fig. 8.1 [118], the projectile ionization cross sections measured for hydrogen-like Au^{78+} ions colliding with carbon atoms are plotted as a function of the beam energy. In the figure, the data are compared with plane-wave Born (PWBA) calculations including corrections for the relativistic wavefunctions and the so-called transverse or magnetic contribution, i.e., the term arising from the alpha operator. Within their uncertainties, the experimental data agree well with this theoretical approach applied.

It is instructive to consider the main parametric dependencies for aymp-totically high collision energies. For the ionization of the target by the projectile, Eq. (8.1) provides the perturbative proportionality of the cross section to the projectile charge as Z_P^2. The target charge enters via the shell size, which is proportional to $1/Z_T$ and gives rise to the target dipole moment responsible for the leading part of ionization. Hence, we obtain an approximate target charge dependence of the cross section as Z_T^{-2}. A more detailed consideration, see e.g., [119, 120, 121, 122, 3] yields the γ dependence. One finds that the ionization cross section is composed of two parts. In the first part, the angular momentum projection μ_j does not change (or changes by more than one unit), in the second part, μ_j changes by just one unit. These contributions have the asymptotic behavior for $\gamma \to \infty$

$$\sigma_{\text{fi}}^{\text{ion}}(\mu_{j_i} \to \mu_{j_i}) \propto Z_P^2 Z_T^{-2} \tag{8.7}$$

and

$$\sigma_{\text{fi}}^{\text{ion}}(\mu_{j_i} \to \mu_{j_i} \pm 1) \propto Z_P^2 Z_T^{-2} \ln \gamma, \tag{8.8}$$

where j_i identifies the initial electron state. The expression (8.7) is mainly due to the Coulomb interaction while (8.8) is exclusively caused by the vector potential. The asymptotic increase of the cross section (8.8) as $(\ln \gamma)$ can be understood by the extension in the transverse direction of the Liénard-Wiechert potential shown in Fig. 5.1 which entails contributions from larger and larger impact parameters as the energy increases. Clearly, the total cross section has the dependence as $\ln \gamma$ resulting from Eq. (8.8). It follows that at very high collision energies, ionization is the dominant atomic process.

Excitation

The initial and final bound states i and f are specified by their quantum numbers $i \equiv \{n_i, \kappa_i, \mu_{j_i}\}$ and $f \equiv \{n_f, \kappa_f, \mu_{j_f}\}$, respectively. In order to obtain the complete cross section, one averages over the initial angular momentum projections μ_{j_i} and sums over the final angular momentum projections μ_{j_f}. In many cases, it is experimentally more practicable to measure the excitation of a highly-charged projectile by the target rather than the excitation of the neutral target by the projectile. In this case, the labels P and T have to be interchanged in the preceding equations.

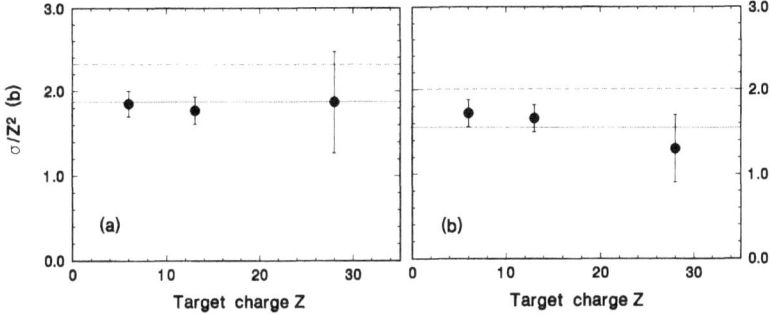

Figure 8.2: Reduced (a) $1s_{1/2} \to 2p_{3/2}$ and (b) $1s_{1/2} \to 2s_{1/2}, 2p_{1/2}$ cross sections (σ/Z_T^2) for Bi^{82+} projectiles of 119 MeV/u impinging on a target with the charge number Z. Theoretical results are obtained *with* (solid line) and *without* (dashed line) the magnetic part of the Liénard-Wiechert interaction. Data from [124].

In contrast to ionization, almost no experimental data are available for projectile excitation in relativistic collisions. However, for a recent review on projectile-electron excitation and loss, see [123]. The first experimental study of projectile K-shell excitation for high-Z ions was reported [124] for the case of hydrogen-like Bi. The excitation process was unambiguously identified by observing the radiative decay of the excited levels to the vacant 1s-shell, in coincidence with projectile ions that did not undergo any charge exchange in the reaction. This allowed one to measure Coulomb excitation of the K-shell electrons in hydrogen-like bismuth ($Z=83$) into specific L-shell sublevels. The measured results indicate an interference between the electric and magnetic parts of the Liénard-Wiechert interaction, most pronounced for excitation to the $2p_{3/2}$ level.

This is illustrated in Fig. 8.2. Here, the reduced cross-section values (σ/Z_T^2) for Ly-α_1 ($1s_{1/2} \to 2p_{3/2}$) and Ly-α_2 ($1s_{1/2} \to 2s_{1/2}, 2p_{1/2}$) excitation are compared with the results of the relativistic first-order perturbation theory, which takes the complete Liénard-Wiechert interaction into account without any further approximations (solid line). For comparison, results are shown obtained by adding the electric and magnetic probabilities *incoherently* (dashed line), that is, ignoring interference terms. All three measurements (for both Ly-α_1 and Ly-α_2) agree with the solid line and disagree with the dashed line. This is contrary to a commonly used assumption that the magnetic part of the interaction always leads to an enhancement of the cross section [121]. Instead, the interference between the electric and the magnetic parts leads to a reduction in the case considered.

We briefly mention some instructive theoretical results for projectile excitation [123]. Since the target atoms are usually neutral, projectile excitation is a many-electron process, which is not in the center of the current presentation. There are two possible modes depending on the involvment of the target. (a) In the so-called "screening mode", the target electrons are considered as inert, so that they only contribute to projectile excitation by partially screening the charge of the target nucleus. (b) The so-called "antiscreening mode" denotes processes, in which also target electrons are excited, so that the term "mutual excitation" would be more appropriate.

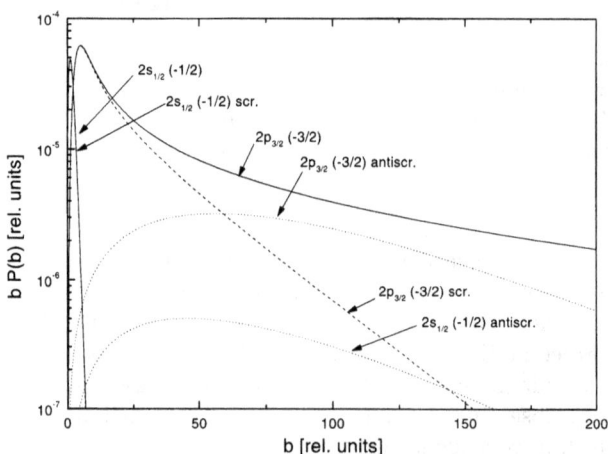

Figure 8.3: Weighted probabilities for projectile excitation in Bi^{82+} collisions with He at $\gamma = 100$. Dashed lines: screening mode, dotted lines: antiscreening mode, solid lines: collisions with bare He^{2+}. The numbers in brackets denote the magnetic quantum numbers of the final states in the Bi^{82+} ion. From [123].

Because of shell size, projectile excitation of specific shells strongly depends on the impact parameter. It turns out that for Bi^{82+} + He collisions with a Lorentz factor of $\gamma = 10$, excitations into the $2s_{1/2}$ and into the $2p_{1/2}$ and $2p_{3/2}$ shells can be well distinguished by their impact-parameter dependence, but the effect of screening is small. However, at $\gamma = 100$, see Fig. 8.3, the situation changes. Expectedly, the influence of screening becomes clearly visible at large impact parameters, and the range of mutual excitation is much larger than for screening, namely about as large as for the unscreened collision.

8.2 Long-range couplings in perturbation theory

In the preceding Section, we have assumed that the initial and final states are undistorted target eigenstates. However, according to the discussion in Sec. 7.4, we should use the exact solution[†] (7.21) of the asymptotic Dirac equation (7.16), that is, one should take into account the asymptotic distortion caused by the residual interaction $-S^2 Z_P e^2/R'$. Inserting the solution (7.21) into Eq. (7.1), one finds that, by construction, the residual interaction term proportional to $1/R'$ has to be subtracted from the interaction occurring in Eq. (7.1), see Eq. (7.24). The reason why one usually disregards the residual interaction term derives from nonrelativistic collisions, where $R = R'$, so that a term proportional to $1/R$ has vanishing matrix elements between orthogonal states.

For relativistic collisions, the situation is somewhat different. Let us examine the modified transition amplitude $\tilde{A}^e$ calculated with phase-distorted states Φ_T^∞ satisfying Coulomb boundary conditions [109]. Since the state-independent phase factors in Eq. (7.21) for initial and final states cancel each other, one has

$$\tilde{A}_{\mathrm{fi}}^e(b) = \mathrm{i}\frac{\gamma e^2 Z_P}{\hbar} \int \mathrm{d}t \, \mathrm{e}^{\frac{\mathrm{i}}{\hbar}(E_\mathrm{f} - E_\mathrm{i})t} \left\langle \mathrm{f} \left| (1 - \beta\alpha_z) \left(\frac{1}{r_P'} - \frac{1}{R'} \right) \right| \mathrm{i} \right\rangle \quad (8.9)$$

instead of (8.1). The additional term $1/R'$ in the perturbation operator with $R' = \sqrt{b^2 + v^2 t'^2}$ depends on the laboratory space coordinates through $t' = \gamma(t - v z_\mathrm{T}/c^2)$. In contrast to the nonrelativistic case, the space integral of the $1/R'$ term does *not* vanish. However, we still have to perform the time integration. Since the integration $\int \mathrm{d}^4 x$ over the complete four-dimensional space is Lorentz-invariant, we may as well evaluate this integral in the projectile frame. In the moving coordinate system, the term $1/R'$ is a constant, and therefore does not contribute to the matrix element between orthogonal (exact) eigenstates, and hence can be discarded. We may now transform back to the laboratory frame to obtain a result *without* the $1/R'$ term, that is, we recover Eq. (8.1). In other words, Eqs. (8.1) and (8.9) are equivalent, and Coulomb boundary conditions are without any influence for direct reactions. It should be emphasized here that the equality of the two treatments is valid *only in first-order perturbation theory* and only if we consider the occupation amplitude for the final state at time $t \to \infty$.

As an illustration, we show in Fig. 8.4 the time evolution of occupation probabilities calculated from Eqs. (8.1) and (8.9), respectively, with the modification that the upper limit of the time integration is taken as t rather than as ∞. In the upper part of the figure, calculated from the

---

[†]An irrelevant constant phase factor depending on the unit of length is discarded.

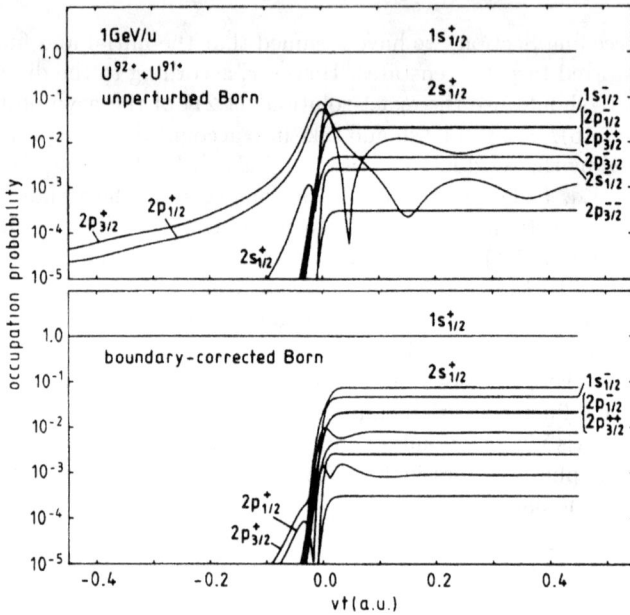

Figure 8.4: Time evolution of the occupation probabilities of target states in U^{92+} + $U^{91+}(1s_{1/2})$ collisions at 1 GeV/u. Calculations are performed in the Born approximation at $b = 0.01$ a.u. (about one K-shell radius). Upper part: unperturbed initial and final states, Eq. (8.1); lower part: boundary-corrected states, Eq. (8.9). The abscissa is the projection of the projectile-target separation on the beam direction. The projections of the angular momenta are indicated by $+, -, ++, --$ for $m_j = +\frac{1}{2}, -\frac{1}{2}, +\frac{3}{2}$, and $-\frac{3}{2}$, respectively. From [109].

finite-time version of Eq. (8.1), we notice the effect of the long-range couplings which lead to a sizable population of $2p_{3/2}$ and $2p_{1/2}$ states in the entrance channel for projectile-target separations as large as 0.4 a.u. (about 37 K-shell radii). On the other hand, in the lower part of the figure, the finite-time version of Eq. (8.9) has been used. Here, according to Eq. (8.9), the long-range couplings have disappeared. For large positive times, the occupation probabilities in the unperturbed and in the boundary-corrected Born approximation are exactly the same, thus verifying the reasoning given above. In a numerical evaluation of the time integrals, one expects that the boundary-corrected version converges much more rapidly.

8.3   Two-center coupled-channel methods

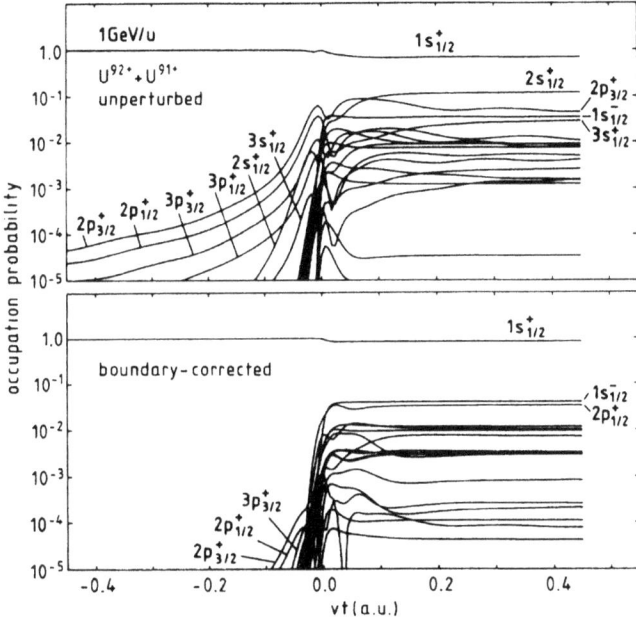

Figure 8.5: Time evolution of the occupation probabilities of target states in $U^{92+} + U^{91+}(1s_{1/2})$ collisions at 1 GeV/u laboratory energy. Calculations are performed with 18 hydrogenic basis states at each of target and projectile and for an impact parameter $b = 0.01$ a.u. The upper set of curves represents results obtained from unperturbed basis states (7.10), while the lower set of curves corresponds to boundary-corrected basis states, Eq. (7.23). For the notation, see Fig. 8.4. From [109].

As has been pointed out in Sec. 7.3, it is necessary for low or intermediate energies to adopt nonperturbative approaches. A well-established method is provided by solving the time-dependent Dirac equation (7.1) directly by inserting an expansion (7.10) or (7.23) for the wavefunction. Within this procedure, the time evolution of the system is explicitly pursued from $t = -\infty$ to $t = \infty$, at least in principle. This implies that long-range couplings, see Sec. 7.5, will have an important effect.

The elimination of the long-range couplings by a suitable gauge trans-

formation   is illustrated in Fig. 8.5. Here, the time evolution during a
U^{92+} + U^{91+} collision at 1 GeV/u is shown for the occupation probabilities of the target states. The impact parameter is 0.01 a.u. or about one
uranium K-shell radius, see Eq. (1.3). Unperturbed basis states (7.10) are
used in the upper part of the figure, while the lower part is obtained for
boundary-corrected basis states (7.23). Here, the long-range couplings visi-
ble for unperturbed basis states have disappeared. The $2p_{3/2}(\frac{1}{2})$, $2p_{1/2}(\frac{1}{2})$,
$3p_{3/2}(\frac{1}{2})$, and $3p_{1/2}(\frac{1}{2})$ states that are excited for an unperturbed basis
already long before the distance of closest approach is reached, begin to be
populated not sooner than about $vt = -10a_K$, if boundary-corrected basis
states are used. Similarly, after the collision, the boundary-corrected basis states attain their asymptotic occupation probabilities at much smaller
values of $z = vt$ than the unperturbed basis states. Most importantly, the
asymptotic values themselves are changed, in sharp contrast to the Born
approximation discussed in Sec. 8.2. A similar behavior has been confirmed
for single-center calculations [109]. Generally speaking, Fig. 8.5 explicitly
illustrates that for an incomplete basis set the coupled-channel equations
(7.12) are not invariant with respect to gauge transformations. It is to
be expected that the more realistic basis states (7.23) will lead to a faster
convergence of the expansion, so that fewer basis states are needed. On
the other hand, the rapid oscillations of the phase factors will render the
numerical evaluation more difficult.

8.4   Calculations on a lattice

The method of solving Eq. (7.1) numerically by finite-difference techniques
in position space or momentum space has been discussed in Secs. 7.6 and
7.7, respectively. The formulation in position space has the drawback that
the grid has to be enclosed in a finite box at the boundaries of which,
that is, when the projectile enters the box, the target can usually not be
regarded as undistorted, at least if the numerically cumbersome long-range
couplings have not been taken into account.

   The formulation in momentum space, on the other hand, while free
from the localization problem, has to cope with the difficulty of an integro-
differential equation and the necessity of taking very small time steps. In
both cases, one has to construct the desired transition amplitude by numerically calculating the overlap of the numerical solution with the relevant
channel functions at large finite times.

   Both methods have been demonstrated but not employed to a large
extent.

Chapter 9

Relativistic electron transfer

The transfer of an electron between target and projectile is one of the basic processes in atomic collision physics. In this Chapter, we consider nonradiative (NRC) or Coulomb capture, while the discussion of another process, namely electron transfer with the simultaneous emission of a photon or "radiative electron capture" (REC), is deferred to Chapter 10.

Similarly as in Chapter 4, we start in Sec. 9.1 with the simplest first-order formulation, the relativistic Oppenheimer-Brinkman-Kramers (OBK) approximation, however, because of similar defects as in the nonrelativistic case, we subsequently introduce the boundary-corrected first-order Born approximation, which takes the long range of the Coulomb potential into account. Analogous to Chapter 4, it turns out in Sec. 9.2 that the relativistic eikonal approximation offers a realistic and yet simple description. Within an exact evaluation, we give examples of charge- and energy dependencies and subsequently furnish an approximate cross-section formula in the form of a very compact analytical expression. In Sec. 9.3, we then turn to the basically exact but numerically expensive description by the two-center coupled-channel method, which allows for a test of the approximate methods discussed before. After this, we are in a position in Sec. 9.4 to compare theoretical and experimental cross sections. Since in all coupled-channel calculations the question arises whether convergence is achieved, we examine in Sec. 9.5 the frame- and basis-set dependence of the theoretical results as a criterion for convergence.

From nonrelativistic capture theory it is known, see Secs. 4.1 and 4.8 and also [2] that for large projectile velocities v the nonradiative charge-changing cross section decreases asymptotically as v^{-12} or E^{-6} in first-order and as v^{-11} or $E^{-11/2}$ in second-order perturbation theory. This dramatic velocity dependence occurs because the overlap in momentum space between the initial and final electron wavefunctions quickly diminishes as the relative velocity increases, see Fig. 4.1. One then might suspect that no measurable cross section may be left at *relativistic* velocities. However, there is a simple qualitative reason why the relativistic cross section decreases less rapidly with the collision energy than the nonrelativistic estimate would indicate: Owing to the Lorentz contraction of the electron space wavefunction in the beam direction, the momentum wavefunctions acquire (Lorentz-) extended tails in this direction and thus give rise to a less rapidly decreasing overlap in momentum space [3]. This expectation is corroborated by the detailed calculations discussed in the following Sections.

9.1 The cross section in first order

As has been shown in Chapter 8 for excitation and ionization, the first-order perturbation theory is a valid approximation, except for very small impact parameters, where the interaction may become extremely strong. For re-arrangement collisions, the situation is more complicated.

The relativistic generalization of the OBK approximation, see Sec. 4.1, has been first worked out in [125, 126], assuming that the initial and final electronic states are undistorted atomic eigenstates φ_i and φ_f of target and projectile, respectively. Strictly speaking, this assumption is not valid since a Coulomb field leads to distortions even at infinite distances. The problem of asymptotic distortions is discussed in Sec. 4.3 for nonrelativistic capture and in Secs. 7.4 and 7.5 for relativistic direct reactions.

Disregarding these problems for a moment, we may adopt the amplitude (7.8) given by perturbation theory and rewrite it more explicitly in the prior form as

$$A_{fi} = i\frac{e^2}{\hbar} \int dt \int d^3 r_T \, \varphi_f'^\dagger(\mathbf{r}_P') \, e^{iE_f t'/\hbar} S \frac{Z_P}{r_P'} \varphi_i(\mathbf{r}_T) \, e^{-iE_i t/\hbar}, \qquad (9.1)$$

where E_i and E_f are the atomic eigenenergies including the electron rest mass, S denotes the spinor transformation (6.9), while φ_i and φ_f' are initial and final bound-state spinor wavefunctions in the target and projectile frame, respectively. At this point, we may convince ourselves that in the nonrelativistic limit (energies ϵ_i, ϵ_f), the phase describing the time oscillation in the projectile frame

$$E_f t' = \gamma E_f\left(t - \frac{v z_T}{c^2}\right) = m_e c^2 t + \epsilon_f t + \tfrac{1}{2} m_e v^2 t - m_e(\mathbf{v}\cdot\mathbf{r}_T) + \cdots$$

$$(9.2)$$

consists of an immaterial contribution due to the electron mass (which is cancelled by the corresponding term from E_i), the nonrelativistic eigenoscillation, and the translation factor, see the expression (2.19). The nonrelativistic limit (4.2) of the transition amplitude (9.1) can be obtained as a special case. The asymptotic behavior of the OBK cross section for very large energies turns out to be

$$\sigma_{\text{OBK}}(1s-1s) \longrightarrow \frac{Z_P^5 Z_T^5}{\gamma} \propto \frac{Z_P^5 Z_T^5}{E} . \tag{9.3}$$

The asymptotic dependence as E^{-1} is characteristic for all relativistic capture theories and has to be compared with the nonrelativistic limit E^{-6}, see (4.11). The charge-state dependence as $Z_{P,T}^5$ reflects the availability of high momentum components in the bound target and projectile wavefunctions, similarly as for the nonrelativistic case.

The exact result for intermediate energies cannot be given in a closed form. An approximate analytical expression for relativistic 1s-1s transitions is given in Eq. (9.12) with (9.13) as a limiting case of the relativistic eikonal approximation, see Sec. 9.2. As has been mentioned above, the slower decrease of the cross section with energy as compared to the nonrelativistic expectation can be traced back to the slower decrease of the overlap of the momentum wavefunctions, see Fig. 4.1, caused by their Lorentz extension as a consequence of the Lorentz contraction in position space.

Similarly as its nonrelativistic counterpart, the relativistic OBK approximation has serious defects: The calculated cross sections are by one order of magnitude too large compared to more realistic calculations and compared to experimental data, with the discrepancy becoming worse if the second order, OBK2, is included [127], see also Table 9.1.

Moreover – and this is the reason – the wavefunctions in entrance and exit channels do not satisfy the correct asymptotic wave equations, Eqs. (7.16) and (7.17). Indeed the long-range nature of the Coulomb interaction does not allow one to use unperturbed atomic wavefunctions even at infinite separations. In analogy to the nonrelativistic case, it is necessary to construct a first-order transition amplitude from the solutions of the asymptotic Dirac equations (7.16) and (7.17). A boundary-corrected first Born approximation for relativistic electron capture has been first introduced in [108]. Since we keep the nonrelativistic acronyms throughout these notes, we here denote it as B1B approximation.

Adopting a formulation in the target inertial frame, the solutions of the asymptotic equations for the initial target and the final projectile states can be written in the forms (7.21) and (7.22) as[†]

$$\Phi_i^\infty(\mathbf{r}_T, t) = e^{-i\nu_P \ln(vR' - v^2 t')} \psi_i(\mathbf{r}_T, t) \qquad \text{for } t \to -\infty \qquad (9.4)$$

and

$$\Phi_f^\infty(\mathbf{r}_T, t) = e^{i\nu_T \ln(vR + v^2 t)} S^{-1} \psi_f'(\mathbf{r}_P', t') \qquad \text{for } t \to +\infty \qquad (9.5)$$

where the Sommerfeld parameters are given by Eq. (7.18), while the internuclear separations R and R' are defined by Eqs. (7.15). The phase factors, which are absent in the OBK approximation, serve to eliminate the asymptotic Liénard-Wiechert interactions caused by the projectile in the laboratory frame [Eq. (7.16)] and by the target in the projectile frame [Eq. (7.17)].

In contrast to the nonrelativistic case, the target wave function (9.4) does not depend on the target time variable t alone but also on the projectile time coordinate t'. Conversely, the projectile wavefunction (9.5) also depends on the target time t. Now, since $t' = \gamma(t - vz_T/c^2)$ and $t = \gamma(t' + vz_P'/c^2)$, this means that the *electronic space coordinates* z_T and z_P' enter in the phase factors of (9.4) and (9.5), respectively. This is in contrast to the nonrelativistic distorted-wave theory, where only the *internuclear coordinates* enter in the distorting phase factor.

9.2   The relativistic eikonal approximation

It is the aim of a distorted-wave treatment of electron transfer to incorporate the effect of the projectile on the target states and the effect of the target on the projectile states. A suggestive ansatz for a more accurate description of the mutual distortion is given by initial and final wavefunctions, each of them being a product of a bound-state wavefunction with respect to one of the partners times a Coulomb distortion factor originating from the other collision partner. This concept is the basis of the "continuum distorted-wave" (CDW) approximation [76], briefly discussed in Sec. 4.5. This approach is successful in nonrelativistic collisions, but problematic to implement for relativistic collisions [128, 129, 3], because in an exact relativistic description, a closed-form continuum wavefunction with a distortion analogous to Eq. (3.41) or (4.27) does not exist and recourse has to be taken to approximate quasirelativistic wavefunctions, see the discussion in [129].

[†]A constant phase factor depending on the unit of length is dropped here and in the following phase factors.

A related but simpler approximation is obtained, if the full Coulomb continuum wavefunction is replaced with its asymptotic form, namely a phase factor depending on the electronic coordinates. In a more general context, this approach is called the eikonal method, see Sec. 4.6 for the nonrelativistic case. The relativistic eikonal approximation has been introduced and worked out for arbitrary initial and final states by Eichler [130]. Since a relativistic symmetrical treatment gives rise to serious problems [128] and misrepresents the experimental data, it is reasonable to adopt the unsymmetrical treatment, in which only the initial *or* the final wavefunction is distorted. To be specific, let us choose the prior form. Following the presentation of the OBK approximation in Eq. (9.1), the eikonal transition amplitude in the prior form is written as

$$A_{\mathrm{fi}} = \mathrm{i}\frac{e^2}{\hbar} \int \mathrm{d}t \int \mathrm{d}^3 r_{\mathrm{T}} \left(\psi_{\mathrm{f}}'(\mathbf{r}_{\mathrm{P}}',t)\right)^{\dagger} S \frac{Z_{\mathrm{P}}}{r_{\mathrm{P}}'} \psi_{\mathrm{i}}(\mathbf{r}_{\mathrm{T}},t) \tag{9.6}$$

with the initial and final wave functions[†]

$$\psi_{\mathrm{i}}(\mathbf{r}_{\mathrm{T}},t) = \varphi_{\mathrm{i}}(\mathbf{r}_{\mathrm{T}})\,\mathrm{e}^{-\mathrm{i}E_{\mathrm{i}}t/\hbar} \tag{9.7}$$

and

$$\psi_{\mathrm{f}}'(\mathbf{r}_{\mathrm{P}}',t') = \varphi_{\mathrm{f}}'(\mathbf{r}_{\mathrm{P}}')\,\mathrm{e}^{-\mathrm{i}E_{\mathrm{f}}t'/\hbar}\,\mathrm{e}^{\mathrm{i}\nu_{\mathrm{T}}'\ln(vr_{\mathrm{T}}+\mathbf{v}\cdot\mathbf{r}_{\mathrm{T}})} \tag{9.8}$$

as a relativistic generalization of (4.30). Here, the unperturbed target and projectile wavefunctions φ_{i} and φ_{f}' are expressed in their respective frames. The Sommerfeld parameter $\nu_{\mathrm{T}}' = Z_{\mathrm{T}}' e^2 / hv$ and the charge number Z_{T}' in the eikonal wavefunction (9.8) are temporarily labeled by a prime in order to provide a unique signature for the electron-target interaction in the *eikonal phase*. This allows one to recover the OBK approximation by simply setting $Z_{\mathrm{T}}' = 0$ or $\nu_{\mathrm{T}}' = 0$.

In the prior version of the eikonal approach, the final wavefunction is phase-distorted by the electron-target interaction. The distorting phase factor $\exp[\mathrm{i}\nu_{\mathrm{T}}' \ln(vr_{\mathrm{T}} + \mathbf{v}\cdot\mathbf{r}_{\mathrm{T}})]$ is the asymptotic form of the Coulomb wavefunction, see Eq. (4.26) for the nonrelativistic case. It may also be represented by an eikonal integral through a relation corresponding to Eq. (4.31). The approximation may also be formulated in the *post* version, which is obtained by inverting the role of target and projectile, that is by replacing Z_{P} by Z_{T} and *vice versa* and by interchanging initial and final states. As the theory is not post-prior symmetric, both versions yield different results.

[†]A constant phase factor depending on the unit of length is dropped in the eikonal phase factor.

Since in an asymmetric theory, the stronger one of the electron-target and electron-projectile interaction should be treated nonperturbatively, the usual prescription, which we denote as Z-criterion, is

if $\quad Z_P \quad < \quad Z_T \quad$ then use the *prior* form

if $\quad Z_P \quad > \quad Z_T \quad$ then use the *post* form. $\qquad$ (9.9)

The exact eikonal cross section resulting from (9.6) can be evaluated numerically, see the tabulation of [131]. For the purpose of illustration, we present in Figs. 9.1 and 9.2 examples for the charge and energy dependence of the cross section. The calculations have been performed adopting the criterion (9.9).

Figure 9.1 exhibits the dependence on target and projectile charges for the leading transition $(1s_{1/2}\text{-}1s_{1/2})$ at a fixed projectile energy of 10 GeV/u. The other cross sections follow a similar pattern. The set of curves reflects the approximate dependence as $Z_T^5 Z_P^5$ of the capture cross section.

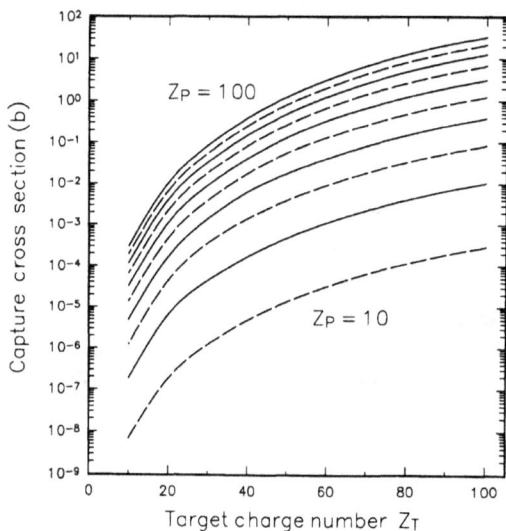

Figure 9.1: Cross sections for $1s_{1/2}\text{-}1s_{1/2}$ electron capture at a projectile energy of 10 GeV/u as a function of the target and projectile charges Z_T and Z_P in steps of 10 units. The cross sections refer to a single target electron captured into a vacant projectile K-shell. They are averaged over the initial and summed over the final angular momentum projections. From [3, 131].

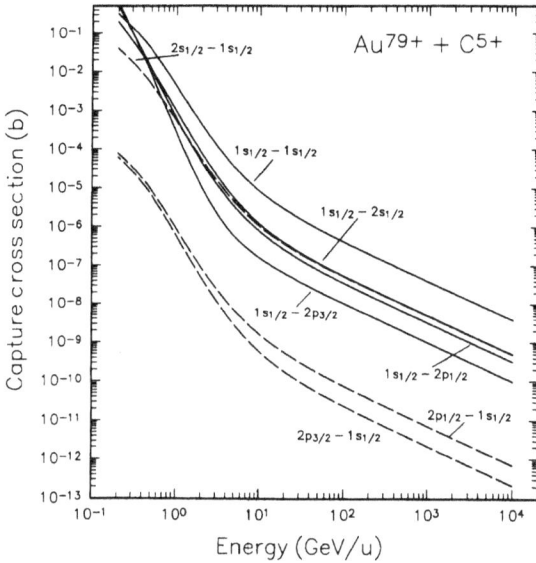

Figure 9.2: Cross sections for electron capture by Au^{79+} ions as a function of the collision energy. See caption of Fig. 9.1. From [3, 131].

Figure 9.2 displays the capture cross sections for completely stripped Au ions impinging on hydrogen-like carbon as a function of energy. The data are averaged over the initial and summed over the final angular-momentum projections. In the state-to-state cross sections presented, either the initial or the final state is a $1s_{1/2}$ state (in order to reduce the number of curves). As a function of the projectile energy, the curves clearly exhibit the transition from an essentially nonrelativistic to a relativistic behavior: at rather low collision energies around 1 GeV/u, the cross section curves still reflect the rapid decrease as E^{-6} known from nonrelativistic collisions, while in the high relativistic energy range above 10 GeV/u, the curves bend over to the E^{-1} dependence. For less asymmetric systems, the spread of curves is much smaller than presented here.

The eikonal cross sections are available in tabular form for more initial and final states and also for uranium projectiles [131]. These tables should furnish a useful guideline for estimating capture cross sections, see Table 9.1 and the comparison with experimental data in Sec. 9.4.

In spite of the existence of these tabulations, it should be useful for an easy application to give an approximate analytical expression. Any relativistic treatment includes two kinds of relativistic effects, namely the relativistic projectile motion, characterized by the parameter γ or, alternatively, by $\delta = [(\gamma - 1)/(\gamma + 1)]^{1/2}$, Eq. (6.10), and the relativistic motion of the electron, characterized by αZ, in its initial and final atomic orbits. Compared to the nonrelativistic case, the relativistic electron motion has two important consequences. (a) The electron orbitals and their binding energies are modified. (b) The electron acquires a Dirac magnetic moment which, in turn, interacts with the induced magnetic field produced by the projectile motion. The occurrence of a Dirac magnetic moment is, of course, independent of the charge Z.

In order to derive an approximate cross-section formula [130] we expand the electronic wavefunctions in powers of $\alpha Z_{P,T} \ll 1$, keeping only the leading terms and at the same time adopting a high-energy approximation for the correction terms in the cross section. We obtain the eikonal cross section by setting $Z'_T = Z_T$ and the OBK cross section by setting $Z'_T = 0$. For convenience, we state the two expressions separately. The 1s-1s eikonal cross section per initial electron summed over the two final spin states is given by

$$
\sigma^{\text{eik}}_{1s-1s} = \frac{2^8 \pi Z_P^5 Z_T^5}{5v^2(Z_T^2 + q_+^2)^5} \frac{\gamma + 1}{2\gamma^2} \frac{\pi \nu_T}{\sinh(\pi \nu_T)} e^{-2\nu_T \arctan(q_+/Z_T)}
$$

$$
\times \left(S_{\text{eik}} + S_{\text{mag}} + S_{\text{orb}} \right) \tag{9.10}
$$

with

$$
S_{\text{eik}} = 1 - \tfrac{5}{4}v^{-1}q_+ + \tfrac{5}{12}v^{-2}q_+^2 + \tfrac{1}{6}v^{-2}Z_T^2
$$

$$
S_{\text{mag}} = -\delta^2 + \tfrac{5}{16}\delta^4 + \tfrac{5}{8}\delta^2\frac{\gamma}{\gamma + 1} + \tfrac{1}{4}\delta^2 v^{-2}Z_T^2 + \tfrac{5}{48}\delta^4 v^{-2}Z_T^2
$$

$$
S_{\text{orb}} = \tfrac{5}{18}\pi\delta\alpha(Z_P + Z_T) - \tfrac{5}{36}\pi\delta^3\alpha(Z_P + Z_T)
$$

$$
- \tfrac{5}{8}\delta\alpha v^{-1}Z_T^2(1 - \tfrac{1}{2}\delta^2) - \tfrac{5}{18}\pi\delta\frac{\gamma}{\gamma + 1}\alpha Z_P
$$

$$
+ \tfrac{5}{28}\pi\delta\left(\frac{\gamma}{\gamma + 1}\right)^2 \alpha Z_P - \tfrac{5}{28}\pi\delta\frac{\gamma}{\gamma + 1}\alpha(Z_P + Z_T - \delta^2 Z_P),
$$

$$
\tag{9.11}
$$

where $q_+ = (E_i - E_f/\gamma)/v$. The expression (9.10) is convenient for estimating capture cross sections and allows one to discuss a few limiting cases.

(a) If we only keep $S_{\rm eik}$ in Eq. (9.10), we obtain the eikonal cross section for relativistic projectiles but nonrelativistic spinless electrons. (b) If, furthermore, we let $\gamma \to 1$ and use the nonrelativistic limit $q_+^{\rm nonrel} = (\epsilon_{\rm i} - \epsilon_{\rm f})/v + \frac{1}{2}v$, see Eq. (4.4), we retrieve the nonrelativistic eikonal cross section (4.33). (c) The term $S_{\rm mag}$ survives for vanishing target charges and tends to zero with decreasing projectile velocity. Therefore, it is interpreted as a magnetic contribution to capture mediated by the interaction between the relativistically induced magnetic field of the projectile and the Dirac magnetic moment of the electron. (d) The quantity $S_{\rm orb}$ is composed of terms that explicitly include $\alpha Z_{\rm P}$ or $\alpha Z_{\rm T}$ and hence is interpreted as a correction term arising from a relativistic modification of the electronic orbitals.

Setting $Z_{\rm T}' = 0$, the approximate OBK cross section [126] is derived as

$$\sigma_{1s-1s}^{\rm OBK} = \frac{2^8 \pi Z_{\rm P}^5 Z_{\rm T}^5}{5v^2 (Z_{\rm T}^2 + q_+^2)^5} \frac{\gamma + 1}{2\gamma^2} \left(1 + S_{\rm mag}^{\rm OBK} + S_{\rm orb}^{\rm OBK}\right) \tag{9.12}$$

with

$$S_{\rm mag}^{\rm OBK} = -\delta^2 + \tfrac{5}{16}\delta^4$$

$$S_{\rm orb}^{\rm OBK} = \tfrac{5}{18}\pi\delta\alpha(Z_{\rm P} + Z_{\rm T}) - \tfrac{5}{36}\pi\delta^3\alpha(Z_{\rm P} + Z_{\rm T}). \tag{9.13}$$

Similarly as in the nonrelativistic case [49], the exactly evaluated relativistic OBK cross sections derived directly from (9.1) considerably overestimate the experimental results, in particular, if the second-order OBK2 is included [127].

For a rough estimate of total cross sections involving higher principal shells, the eikonal formula (9.10) is extended in the following way: Starting from the observation that the cross section scales with Z/n where n is the principal quantum number of the shell in question, one may formulate an approximate scaling rule [132]. Approximate relativistic capture cross sections averaged over subshells within arbitrary initial and final principal shells specified by their principal quantum numbers $n_{\rm T}$ and $n_{\rm P}$, respectively, can be obtained from Eq. (9.10) by replacing, in the prior form, as

$$Z_{\rm T} \to \frac{Z_{\rm T}}{n_{\rm T}} \qquad \text{and correspondingly,} \qquad Z_{\rm P} \to \frac{Z_{\rm P}}{n_{\rm P}}. \tag{9.14}$$

To obtain the post form, target and projectile as well as initial and final states have to be interchanged. If this scaling rule is adopted, the decision between the prior and the post form will be automatically based on the Z/n-criterion following from the rule (9.9) by substituting the replacement (9.14).

9.3   Two-center coupled-channel methods

The application of two-center coupled-channel methods has been discussed
in Sec. 7.3 in a general context. We here summarize the main points.
The time-dependent Dirac equation (7.1) for relativistic ion-atom collisions is solved by using the expansion (7.10) of the time-dependent wave-
function $\Psi(\mathbf{r}_{\mathrm{T}}, t)$ in terms of target and projectile eigenstates $\psi_k(\mathbf{r}_{\mathrm{T}}, t)$
and $\psi_{k'}(\mathbf{r}'_{\mathrm{P}}, t')$, respectively. The expansion may also include wave packets
(7.11) to represent the continuum. These methods apply equally for charge
transfer.

However, according to the discussion in Sec. 7.5, it is more appropriate
to use a set of basis states that satisfy Coulomb boundary conditions (9.4)
and (9.5) [109]. We therefore replace Eq. (7.10) by the basis expansion
(7.23). Indeed, when we insert (7.23) into the time-dependent Dirac equation (7.1), we find that on the right-hand side, the time differentiation of
the logarithmic phase factors yields additional terms so that for the perturbing potential we obtain the replacement (7.24). These replacements
have to be substituted into the potential matrix elements (7.14) and into
the coupled equations (7.12). The phase factors themselves drop out in
the one-center integrals since they are state-independent, but they remain
in the two-center integrals. We denote the expansion (7.23) as "boundary
corrected" and, in contrast, (7.10) as "unperturbed" since the asymptotic
Coulomb distortion is not accounted for in the basis set. The evaluation of
the time-dependent equations (7.12) proceeds as in Secs. 7.3 and 7.5. The
overlap and potential matrix elements have to be calculated or interpolated
at each time step. This requires most of the computing time. While in the
corresponding nonrelativistic calculations the internuclear line is a useful
symmetry axis reducing the numerical effort, in relativistic collisions, the
z-axis is singled out by the Lorentz transformation and hence destroys the
axial symmetry.

9.4   Theoretical and experimental cross sections

We now have at our disposal several theories, both perturbative and nonperturbative approaches. It is worthwhile to compare their predictions for
capture cross sections among each other and also with existing experimental data. For a comparison between the theories discussed above, we use a
$U^{92+} + U^{91+}$ collision as a test system. For charge transfer, a comparison
is presented in Table 9.1. In addition to B1B and OBK cross sections, see
Sec. 9.1, the prior and post versions of the relativistic eikonal approximation, see Sec. 9.2, are considered. All of them are evaluated numerically

Table 9.1: Comparison of theoretical cross sections per electron (in barn) for 1s capture in collisions of $U^{92+} + U^{91+}$ at 500 MeV/u. CC: 36-state (per center) coupled-channel calculations [107]; B1B: relativistic boundary-corrected Born approximation [108], EA prior and EA post: prior and post form eikonal approximation [130]; OBK: relativistic OBK approximation. The numbers in square brackets give the powers of 10 multiplying the preceding number. From [107]. The table also includes the ratio of spin-flip to non-spin-flip cross sections.

final shell	CC	B1B	EA prior	EA post	OBK
$1s_{1/2}$	4.12[3]	3.39[3]	4.14[3]	4.14[3]	4.91[4]
$2s_{1/2}$	1.06[3]	5.13[2]	7.62[2]	5.67[2]	5.28[3]
$2p_{1/2}$	3.91[2]	5.95[2]	1.35[3]	4.44[2]	9.02[3]
$2p_{3/2}$	2.31[2]	2.74[2]	1.24[3]	1.07[2]	7.61[3]
$3s_{1/2}$	3.28[2]	1.51[2]	2.30[2]	1.58[2]	1.41[3]
$3p_{1/2}$	1.03[2]	1.75[2]	4.37[2]	1.30[2]	2.60[3]
$3p_{3/2}$	7.25[1]	9.15[1]	4.62[2]	3.61[1]	2.56[3]
sum	6.31[3]	5.19[3]	8.62[3]	5.58[3]	7.76[4]
$\sigma^{\mathrm{flip}}/\sigma^{\mathrm{non}}$	5.73[-2]	3.41[-2]	5.84[-2]	5.83[-2]	7.30[-3]

without assuming the αZ expansion underlying Eq. (9.10). We observe that in most cases the B1B cross sections have a magnitude intermediate between the prior and post form eikonal values. For some final subshells, in particular for $1s_{1/2}$, the B1B results are close to the coupled-channel results, for others they deviate by as much as a factor of two. The total K-shell capture cross sections are all in fair agreement with one another except for those of the OBK theory, which systematically yields cross sections that are by one order of magnitude too high.

It is desirable to compare calculated cross sections with experimental values. Total cross sections have been measured for a variety of projectile-target combinations. Figure 9.3 shows a comparison between measured

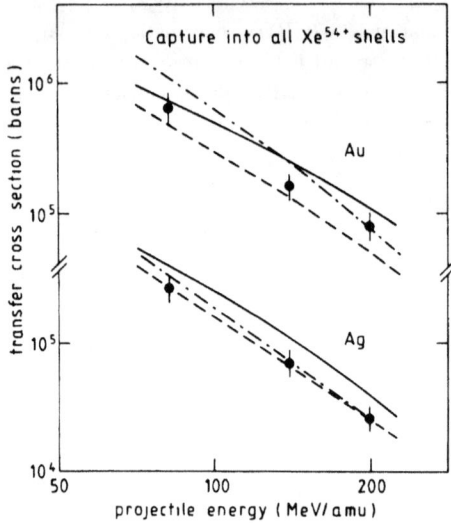

Figure 9.3: Cross sections for charge transfer in Xe^{54+} + Ag and Au collisions as a function of the laboratory projectile energy. The error bars include estimated 20% systematic errors. The solid lines represent the results of coupled-channel calculations [107]. The dashed and dash-dotted lines use the eikonal approximation with the Z [Eq. (9.9)] and Z/n [Eq. (9.14)] criteria, respectively. From [107, 3].

[132] and total calculated eikonal and coupled-channel cross sections for Xe^{54+} projectiles on Ag and Au .

Coupled-channel calculations are available for Xe^{54+} on Ag and Au at collision energies of 82, 140, and 197 MeV/u. Since the coupled-channel theory is constructed for a single-electron target, additional assumptions have to be made for multielectron targets. Since transitions between occupied target shells cannot occur, the contribution of each target subshell has been calculated separately [107] by including only this specific target state in the two-center basis and by blocking all other occupied target states. For the bare projectile, all 28 states of the K-, L-, and M-shells are coupled. One after the other, all target subshells up to $2p_{3/2}$ for Ag and up to $3p_{3/2}$ for Au have been individually considered using effective charges $Z^\star = Z_T$ for the K-shell and, according to the Slater rules, $Z^\star = Z_T - 4.15$ for the

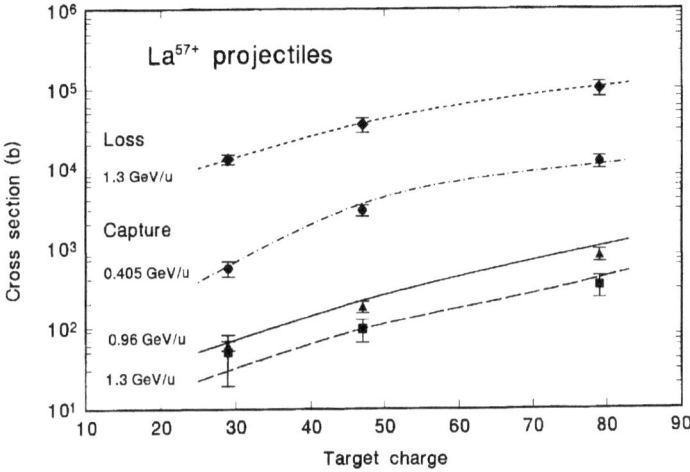

Figure 9.4: Measured total electron-capture cross sections for 0.405, 0.96, and 1.3 GeV/u La^{57+} projectiles and electron loss cross sections for 1.3 GeV/u La^{56+}. The theoretical lines for capture are from the eikonal theory [130, 131] and include contributions from radiative and nonradiative capture. From [133].

L-shell and $Z^\star = Z_{\text{T}} - 11.0$ for the $3s$ and $3p$ shells. While coupled-channel calculations are in reasonable agreement with the data for Au, they overestimate the cross section for Ag by 20–40 %, while the eikonal cross sections are close to the experiment.

Figure 9.4 shows experimental capture cross sections for La^{57+} projectiles impinging on Cu, Ag, and Au targets at 0.405, 0.96, and 1.3 GeV/u [133]. The theoretical cross sections represent the sum of nonradiative capture calculated in the prior form of the eikonal approximation according to [130, 131] and of radiative electron capture, see Chapter 10, calculated [133] from tabulated photoelectric cross sections. An earlier comparison between experimental and theoretical capture data can be found in [132, 3]. It is noted that within the experimental accuracy the agreement is satisfactory.

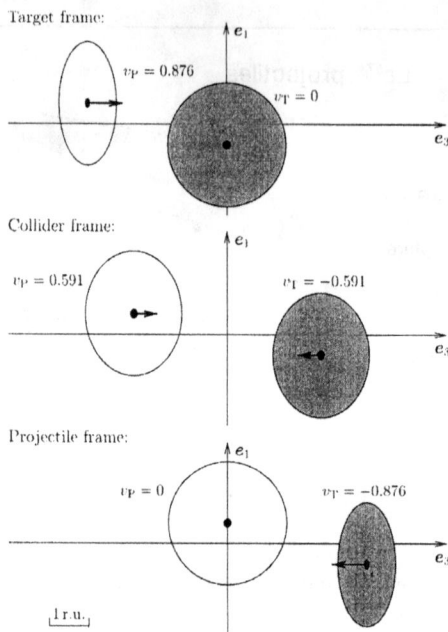

Figure 9.5: Three different frames of reference are shown. The Lorentz contraction of moving spheres corresponds to collision energies of 1.0 GeV/u in all cases. From [134].

9.5   Frame- and basis-set dependence

The cross sections discussed so far, have always been calculated in the target system. A relativistically rigorous treatment is independent of the Lorentz frame in which it is formulated. Calculated cross sections must hence be Lorentz-invariant. The two-center coupled-channel formulation described in Secs. 7.3 and 7.5 is exact provided the basis set used is complete. In reality, this is not the case, hence we may expect a dependence on the Lorentz frame. The frame dependence has been investigated in detail by T. Brunne in his Doctoral Thesis [134]. The goal was to use the frame dependence as an indicator for the completeness or incompleteness of a given basis set and hence for the convergence. As an illustration, Fig. 9.5 shows schematically three different Lorentz frames which are frequently used, all of them for a collision energy of 1.0 Gev/u: The target frame, in

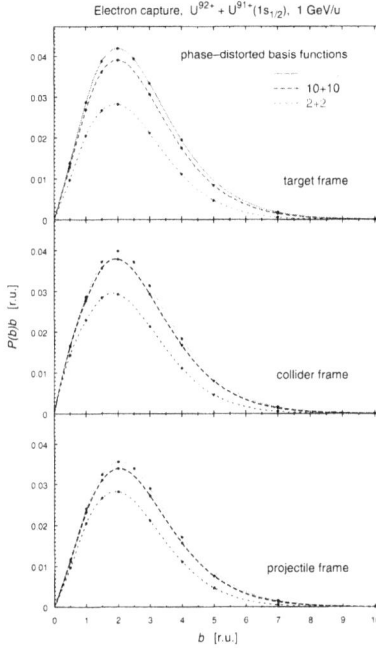

Figure 9.6: Weighted total capture probabilities $P(b)\,b$ obtained from coupled-channel calculations with phase-distorted basis functions according to Eq. (7.23). The initial state is $1s_{1/2}$ and the bases consist of the 2, 10, and 28 lowest bound states at each nucleus. From [134].

which the target is at rest and the projectile is moving, the collider system, in which target and projectile are moving in opposite directions, and finally the projectile system, in which the projectile is at rest and the target is moving.

Of course, there is a continuum of other ways to distribute the energy between target and projectile. A parameter convenient for characterizing different reference frames is

$$\xi = \frac{\zeta_T + \zeta_P}{\zeta_T - \zeta_P}\,, \tag{9.15}$$

where ζ is the rapidity defined in Eq. (5.10). With this convention, $\xi = 0$ corresponds to the collider frame, $\xi = -1$ to the target frame and $\xi = 1$ to the projectile frame.

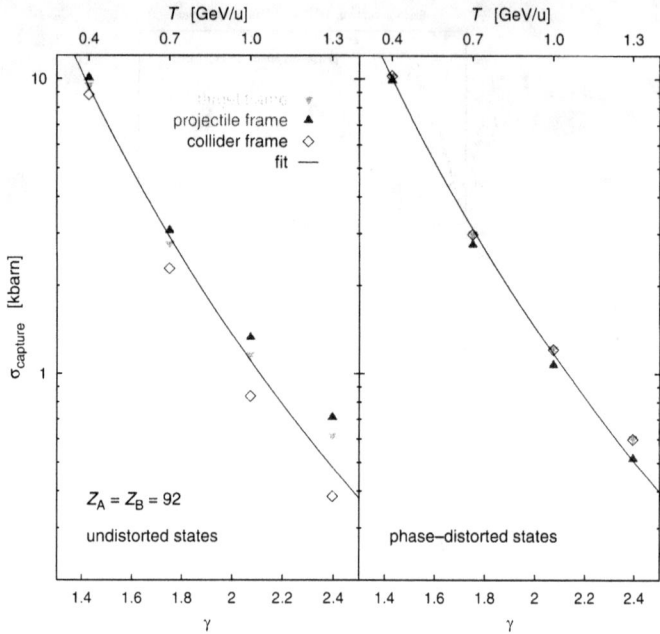

Figure 9.7: Collision-energy dependence of cross sections for electron capture by U^{92+} from U^{91+}(1s) for three different coordinate frames and for undistorted and phase-distorted basis states. From [134].

Calculations have been performed within the two-center coupled-channel method for various basis sizes, N_a at the one partner and N_b at the other partner, the combination denoted as $N_a + N_b$. A comparison has also been made for undistorted basis sets according to Eq. (7.10) and phase-distorted basis sets, taking into account Coulomb boundary conditions, according to Eq. (7.23). As an example, Fig. 9.6 gives the impact-parameter dependence (in relativistic units) of the weighted capture probability $P(b)b$ for $U^{92+} + U^{91+}$ collisions at 1.0 GeV/u. Here the expansion (7.23) in terms of phase-distorted basis states has been used. Quite generally, the maximum of the weighted capture probability occurs at $b \approx 2\lambda_c$, which corresponds to about 1.5 K-shell radii, and rapidly decreases for larger impact parameters. It is seen that an expansion with only two $1s_{1/2}$ states (2+2) at each center yield an approximately correct capture probability. The increase of the basis from the lowest (10+10) to (28+28)

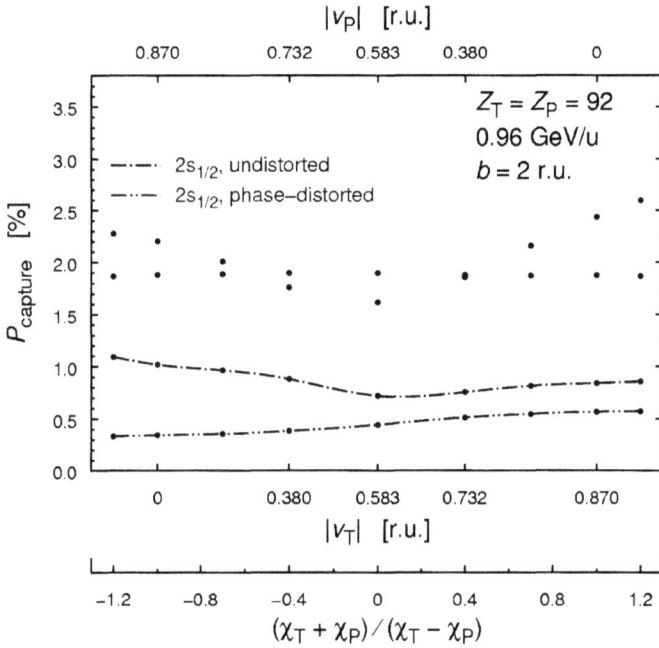

Figure 9.8: Frame dependence of capture probabilities in percent for U^{92+} – U^{92+} collisions at 0.96 GeV/u and impact parameter $b = 2$ r.u. From [134].

states does not change the result very much. Also the differences between the results for the three coordinate systems are less than 20 percent. The calculations for undistorted basis states yield qualitatively similar results. However, it is noted that the differences between results obtained for different Lorentz frames are smaller for calculations with phase-distorted basis functions than for undistorted functions. This suggests that the frame dependence of the capture calculations (as an indicator of convergence with regard to the number of basis states included) is reduced by using phase-distorted wavefunctions. Nevertheless, it is remarkable that the difference between results obtained with undistorted and phase-distorted wavefunctions is much smaller than the corresponding difference for perturbative results, compare the columns B1B and OBK in Table 9.1.

Figure 9.7 displays total cross sections for electron capture of U^{92+} from $U^{91+}(1s)$ calculated from the capture probability $P(b)$ according to

Eq. (2.5). We show both the results obtained with unperturbed and with phase-distorted basis functions. While the energy dependence in both cases is very similar, it is again seen that the frame dependence is much smaller, when phase-distorted basis functions are used. This once more suggests that the convergence properties of phase-distorted wavefunctions are better than those of undistorted wavefunctions.

Finally, Fig. 9.8 exhibits the frame dependence of capture probabilities for a whole continuum of reference frames characterized by the parameter ξ defined in Eq. (9.15). In this case, in addition to 10 bound-state basis functions for each center, 24 wave-packet basis functions (7.11) for each center have been included with $\kappa = \pm 1$, $E_k = \pm 1.15$, ± 1.45, ± 1.75 and $\Delta E_k = 0.3$, (all in relativistic units, r.u.), half of them with positive energy and half with negative energy. The radial wavefunctions of these wave packets are approximately localized within a sphere of 200 r.u. Once again it is seen that calculations with phase-distorted basis states (7.21) and (7.22) exhibit much less dependence on the reference frame than undistorted basis states. As expected, this suggests that the former ones are more physical states and lead to a faster convergence.

Chapter 10

Radiative electron capture (REC)

In Chapters 4 and 9, we discuss Coulomb or nonradiative electron capture (NRC). In the nonrelativistic energy regime, the cross section for this process falls off asymptotically with increasing projectile velocity v as v^{-12} (or as v^{-11} in second order). This rapid decrease is mainly caused by the requirement that a given momentum component in the initial electronic wavefunction has to find its counterpart in the final momentum wave function displaced by the momentum $m_e v$ of an electron travelling with the speed of the projectile, see Fig. 4.1. If, however, the electron transfer is accompanied by the emission of electromagnetic radiation, the emitted photon acts as a third body carrying away energy and momentum released by the formation of the final bound state. Hence, the condition of momentum matching will be relaxed so that the cross section for radiative electron capture (REC) falls off only as v^{-5} for high nonrelativistic projectile velocities.

For free electrons, capture cannot take place at all without the emission of photons owing to energy and momentum conservation. This means, qualitatively, that electrons loosely bound in low-Z target atoms or in outer shells are more likely to be captured *with* photon emission than *without*. Hence for low-Z target atoms at high projectile energies (e.g., for $Z_T = 1$ with $E_P \geq 10$ MeV/u), the REC cross section exceeds the cross section for Coulomb capture. From this point of view, the REC mechanism deserves particular attention. For a summary, see [135].

This situation is illustrated in Fig. 10.1. In the extreme relativistic regime, nonradiative as well as radiative electron capture decrease as $1/\gamma$. The asymptotic behavior of the cross section for nonradiative electron

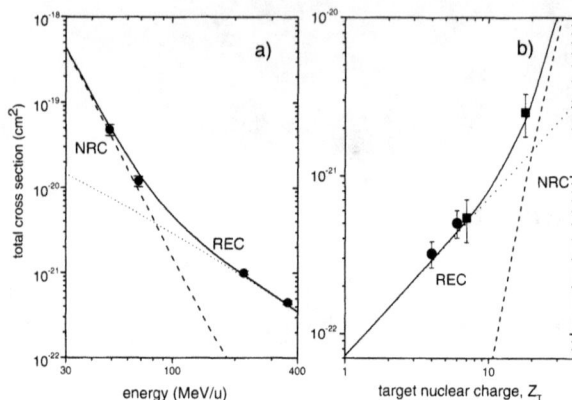

Figure 10.1: (a): Total electron-capture cross sections for U^{92+} on a N_2 target versus projectile energy. The dotted line represents the result of the eikonal approach for the NRC process [130]. The dashed line gives the prediction obtained for REC within the dipole approximation. The solid line refers to the sum of both predictions. (b): Total electron-capture cross sections for bare U^{92+} at 295 MeV/u colliding with gaseous targets and with solid targets. From [135].

transfer is given by Eq. (9.3). If, in radiative electron capture, the electron transfer is accompanied by the emission of electromagnetic radiation, the decrease of the cross section in the asymptotic energy regime has the parametric dependence

$$\sigma^{\mathrm{REC}} \propto Z_{\mathrm{P}}^5 Z_{\mathrm{T}} \frac{1}{E}, \tag{10.1}$$

where the linear dependence on Z_{T} reflects the number of electrons available for capture in a neutral target and Z_{P}^5 has the same origin as in Eq. (9.3).

In this Chapter, we start in Sec. 10.1 with an intuitively suggestive approach, which treats REC within the impulse approximation, that is, as radiative recombination modified by the momentum distribution of the electrons in the target. As the inverse of radiative recombination, we subsequently present in Sec. 10.2 the nonrelativistic Born approximation for K-shell photoionization providing a simple access to qualitative features of REC. In Sec. 10.3, the Stobbe formula then takes into account the Coulomb deflection of nonrelativistic electrons. However, for high charges and high energies, one needs exact relativistic calculations, presented in Sec. 10.4.

We discuss differential cross sections, in particular magnetic spin-flip contributions, and, finally, systematics of total cross sections in Sec. 10.5. In order to gain more insight into the reaction dynamics, alignment of an intermediate state and photon polarization is examined in Sec. 10.6. As a new process, the conversion of a virtual REC photon into an electron-positron pair, is briefly discussed in Sec. 10.7 .

10.1 The impulse approximation

A loosely bound target electron may be considered as approximately free in a high-energy collision. In this limit, REC is identical to radiative recombination (RR), in which an electron initially moving with the velocity $-\mathbf{v}$ in the projectile frame is captured into a bound state of the projectile with the simultaneous emission of a photon of energy $\hbar\omega'$ and wave number $\mathbf{k}'$. Here and in the following, we denote quantities in the projectile frame with a prime and adopt relativistic kinematics for the projectile motion. This situation is schematically illustrated in Fig. 10.2.

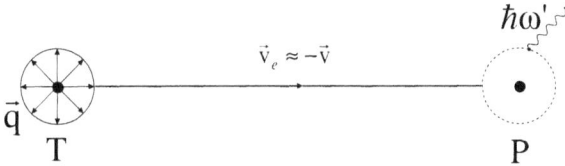

Figure 10.2: Schematic illustration of a target atom T with an internal momentum distribution $\mathbf{q}$ moving towards the projectile. One of the target electrons is captured by the projectile P with the simultaneous emission of a photon $\hbar\omega'$.

Following the seminal paper of Kleber and Jakubassa [136], we now turn to a brief preliminary outline of the treatment. We assume that the cross section σ_{RR} for radiative recombination is known, since it is essentially the inverse of the atomic photoelectric effect, see Fig. 10.3. It is then natural to refer all momenta to the projectile frame. If the target electron has the momentum $\mathbf{q}$ with respect to the target nucleus and the projectile momentum is characterized by the Lorentz factor $\gamma = (1 - \beta^2)^{-1/2}$ with $\beta = v/c$, its momentum in the projectile frame $\mathbf{q}'$ is obtained from relativistic kinematics, see Sec. 5.2. If the momentum $\gamma m_e v$ of an electron travelling with the relative speed v of the target towards the projectile is large compared to the electron momentum q in a low-Z target atom, one

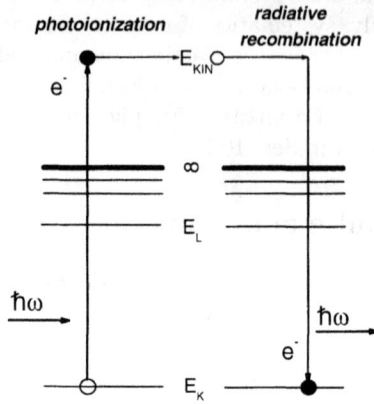

Figure 10.3: Schematic sketch of the time-reversed processes: the photoeffect (PI) and the radiative recombination (RR).

may use the impulse approximation [136] to write the double-differential REC cross section in the projectile frame (primed coordinates) as

$$\frac{d^2 \sigma_{\mathrm{REC}}}{d\Omega' \, d(\hbar\omega')} = \int d^3q \, \frac{d\sigma_{\mathrm{RR}}(q')}{d\Omega'} \, |\tilde{\varphi}_{\mathrm{i}}(\mathbf{q})|^2 \, \delta(\hbar\omega' + E'_{\mathrm{f}} - E'_{\mathrm{i}}), \qquad (10.2)$$

where

$$\tilde{\varphi}_{\mathrm{i}}(\mathbf{q}) = (2\pi\hbar)^{-3/2} \int d^3q \, \varphi_{\mathrm{i}}(\mathbf{r}) \, e^{-i\mathbf{q}\cdot\mathbf{r}/\hbar} \qquad (10.3)$$

is the Fourier transform of the initial electronic target wavefunction. The delta function expresses the energy conservation between the final electronic energy E'_{f} and the photon energy $\hbar\omega'$ in the projectile frame on the one hand and the initial electronic energy E'_{i} (also in the projectile frame) on the other hand. According to the relativistic kinematics discussed in Sec. 5.2, see Eq. (5.15), these energies are written as

$$\begin{aligned} E'_{\mathrm{i}} &= \gamma(E_{\mathrm{i}} - vq_z) \\ &= \gamma m_e c^2 - \gamma|\epsilon_{\mathrm{i}}| - \gamma v q_z \end{aligned} \qquad (10.4)$$

and

$$E'_{\mathrm{f}} = m_e c^2 - |\epsilon'_{\mathrm{f}}|. \qquad (10.5)$$

Here $|\epsilon_i|$ and $|\epsilon'_f|$ are the binding energies of initial target and final projectile states. Inserting the expressions (10.4) and (10.5) into Eq. (10.2), we have

$$\frac{d^2\sigma_{REC}}{d\Omega' \, d(\hbar\omega')} = \int d^3q \, \frac{d\sigma_{RR}(q')}{d\Omega'} \, |\tilde{\varphi}_i(\mathbf{q})|^2 \, \delta(\hbar\omega' + E'_f - \gamma E_i + \gamma v q_z)$$

$$= \int d^3q \, \frac{d\sigma_{RR}(q')}{d\Omega'} \, |\tilde{\varphi}_i(\mathbf{q})|^2$$

$$\times \, \delta(\hbar\omega' - |\epsilon'_f| + \gamma|\epsilon_i| - T_e + \gamma v q_z), \tag{10.6}$$

where $T_e = m_e c^2 (\gamma - 1)$ is the kinetic energy of an electron travelling with the same speed as the projectile. Since the initial momentum distribution of the target electron in the target frame is peaked around $\mathbf{q} \approx 0$ and since we usually consider $Z_T \ll Z_P$, the recombination cross section varies slowly and can be taken outside the integral over the transverse momentum. In this approximation, we get

$$\frac{d\sigma^2_{REC}}{d\Omega' \, d(\hbar\omega')} = \frac{1}{\gamma v} \left(\frac{d\sigma_{RR}}{d\Omega'} \right)_{q'_z = \gamma(-m_e v + q_z)} \, \mathcal{J}_i(q_z), \tag{10.7}$$

where

$$\mathcal{J}_i(q_z) = \int d^2q_\perp \, |\tilde{\varphi}_i(\mathbf{q})|^2 \tag{10.8}$$

is the Compton profile. The integration extends over the transverse momentum $\mathbf{q}_\perp$. In Eq. (10.7), the longitudinal momentum q_z is fixed by ω', so that the shape of the photon spectrum

$$\hbar\omega' = |\epsilon'_f| - \gamma|\epsilon_i| + T_e - \gamma v q_z \tag{10.9}$$

is determined by the spread of the longitudinal momentum q_z around its mean value $q_z = 0$, which yields the peak energy

$$\hbar\omega'_0 = \gamma E_i - E'_f = |\epsilon'_f| - \gamma|\epsilon_i| + T_e. \tag{10.10}$$

The term $\gamma v q_z$ in Eq. (10.9) describes the Doppler broadening through the Compton profile. However, since the cross section σ_{RR} decreases with increasing x-ray energy, the actually measured peak is usually slightly shifted to lower energy.

10.2 Some basics: Born approximation for K-shell photoionization and REC

Following Sec. 10.1, we describe radiative recombination (RR) as the inverse of the atomic photoelectric effect: a continuum electron is captured by the projectile with the simultaneous emission of a photon. A schematic representation of the two processes is given in Fig. 10.3. The original three-body problem is reduced to a two-body problem, and, for the REC process, the target merely enters by providing an initial momentum distribution of the continuum electrons as seen from the projectile nucleus. The initial momentum spread of the electrons bound in the target atom leads to a broadening of the photon resonance, but can in many cases be neglected in calculating differential and total cross sections. As a first step, we therefore have to calculate the photoelectric effect.

The simplest case of photoionization is represented by the absorption of a photon by a nonrelativistic K-electron. For illustrative purposes, we temporarily adopt the plane-wave Born approximation, assuming that the energy of the incident light quantum is large compared to the ionization energy $I = |\epsilon_i|$ of the K-electron. According to the energy balance $T_e = \hbar\omega - |\epsilon_i|$ this also means that

$$T_e \gg I = \frac{Ze^2}{2a_0} \quad \text{or} \quad \nu = \frac{e^2 Z}{\hbar v} \ll 1, \tag{10.11}$$

where $a_0 = \hbar^2/m_e e^2$ is the Bohr radius of the hydrogen atom and ν is the Sommerfeld parameter. Equation (10.11) is identical with the condition for the validity of the Born approximation. This means that we can approximately replace the final electron wavefunction φ_f by a plane wave $\varphi_f(\mathbf{r}) = e^{i\mathbf{k}_f \cdot \mathbf{r}}$, with $\mathbf{k}_f = \mathbf{p}_f/\hbar$ being the electron wave number in the final state. The nonrelativistic initial wavefunction for a hydrogen-like K-shell electron is $\varphi_i(\mathbf{r}) \propto e^{-Zr/a_0}$. The transition is mediated by the electromagnetic interaction $\mathbf{j} \cdot \mathbf{A} \propto \mathbf{p}_{op} \cdot \hat{\mathbf{u}} e^{i\mathbf{k}\cdot\mathbf{r}}$, where $\mathbf{p}_{op} = \frac{\hbar}{i}\nabla$ is the nonrelativistic current operator and $\hat{\mathbf{u}} e^{i\mathbf{k}\cdot\mathbf{r}}$ is the photon wavefunction with the unit vector $\hat{\mathbf{u}}$ for the polarization and the wave vector $\mathbf{k}$. If we act with $\mathbf{p}_{op}$ to the left on $\exp(i\mathbf{k}_f \cdot \mathbf{r})$, and combine the exponentials by abbreviating $\mathbf{k} - \mathbf{k}_f = \mathbf{q}$, we obtain the interaction matrix element as

$$M_{fi} = \int \varphi_f^*(\mathbf{r}) \, \mathbf{p}_{op} \cdot \hat{\mathbf{u}} e^{i\mathbf{k}\cdot\mathbf{r}} \, \varphi_i(\mathbf{r}) \, d^3r = (\mathbf{k}_f \cdot \hat{\mathbf{u}})\hbar \int e^{i\mathbf{q}\cdot\mathbf{r}} \, \varphi_i(\mathbf{r}) \, d^3r, \tag{10.12}$$

which is proportional to the Fourier transform of the initial wave function. Choosing the photon direction as the z-axis and the direction $\hat{\mathbf{u}}$ of the polarization as the x-axis, we have $\mathbf{k}_f \cdot \hat{\mathbf{u}} = k_f \sin\theta \cos\varphi$, where θ is the angle between $\mathbf{k}_f$ and the photon direction and φ the angle between the projection of $\mathbf{k}_f$ onto the x-y plane and the x-axis. Upon inserting the

Fourier transform of the bound-state wavefunction [39] into Eq. (10.12), we obtain [137]

$$|M_{fi}|^2 \propto \frac{\sin^2\theta\cos^2\varphi}{(\alpha^2 Z^2 + q^2 \lambda_c^2)^4} Z^5 \propto \frac{\sin^2\theta\cos^2\varphi}{(1 - \beta\cos\theta)^4} Z^5, \qquad (10.13)$$

which shows the same denominator as the classical expression (1.21). Here, $\lambda_c = \hbar/m_e c$ is the Compton wave length of the electron and $\beta = v/c = p_f/m_e c = k_f \lambda_c$. For the last relation in Eq. (10.13), we have used the energy conservation, i.e. $k\lambda_c = (\alpha^2 + k_f^2 \lambda_c^2)/2$ and $\hbar\omega \ll m_e c^2$. Collecting coefficients and averaging over photon polarizations, one derives [138] the final result in the moving projectile system (denoted by a prime) as

$$\frac{d\sigma_{ph}(\theta')}{d\Omega'} = 2\sqrt{2}\,\alpha^8 Z^5 \left(\frac{m_e c^2}{\hbar\omega}\right)^{7/2} a_0^2 \frac{\sin^2\theta'}{(1 - \beta\cos\theta')^4}. \qquad (10.14)$$

No electrons are emitted in the direction of **k**. This is a direct consequence of angular momentum conservation, which forbids forward or backward emission, since in the nonrelativistic absence of electron spin, the electron cannot carry away the photon spin at $\theta = 0$ or $\theta = \pi$. We come back to this point in Sec. 10.5. The denominator in Eq. (10.14) originates from the "retardation effect", which consists in keeping the full exponential $e^{i\mathbf{k}\cdot\mathbf{r}}$ in the photon wavefunction in Eq. (10.12) and leads to a tilting of the angular distribution in the forward direction. In the relativistic case, see Sec. 10.4, the maximum is strongly displaced towards forward angles.

If retardation is neglected, $\exp(i\mathbf{k}\cdot\mathbf{r}) \to 1$, one obtains the dipole approximation. In this case, no angle enters into the denominator of Eq. (10.14), and the differential cross section is given by a pure $\sin^2\theta$ distribution. For unpolarized primary light we have

$$\frac{d\sigma_{ph}^{dipole}}{d\Omega'} \propto \sin^2\theta'. \qquad (10.15)$$

From the accurate cross section (10.14) for the photoelectric effect in the presence of retardation, one derives the cross section for the inverse process of radiative recombination with the aid of the principle of detailed balance. In this case, the angle θ' is measured with respect to the beam direction, so that $\theta'_{ph} \to \pi - \theta'$ or $\cos\theta'_{ph} \to -\cos\theta'$. As a consequence,

$$\frac{d\sigma_{RR}(\theta')}{d\Omega'} \propto \frac{\sin^2\theta'}{(1 + \beta\cos\theta')^4} \qquad (10.16)$$

has a maximum shifted towards backwards angles in the projectile frame. If we now Lorentz transform Eq. (10.16) to the laboratory system with the aid of Eqs. (5.19) and (5.20), that is

$$\cos\theta' = \frac{\cos\theta - \beta}{1 - \beta\cos\theta} \quad \text{and} \quad \frac{d\Omega'}{d\Omega} = \frac{1}{\gamma^2(1 - \beta\cos\theta)^2}, \tag{10.17}$$

we obtain

$$\frac{d\sigma_{RR}}{d\Omega} \propto \sin^2\theta, \tag{10.18}$$

that is, a pure $\sin^2\theta$ distribution in the *laboratory frame*. This peculiar cancellation between the effects of the retardation, i.e. of higher multipoles (leading to a deviation from a $\sin^2\theta$ distribution) and the Lorentz transformation to the laboratory system has been first observed by Spindler [139]. It will be shown in Sec. 10.4 that this general behavior is still approximately valid for rather high projectile energies, as can be seen in Fig. 10.6.

10.3   The Stobbe formula for K-shell photoionization and REC

The Born approximation described in Sec. 10.2 is not adequate for larger values of Z or if the photon energy $\hbar\omega$ is so small that the energy of the ejected electron is of the same order of magnitude as the ionization energy $I = |\epsilon_i|$, that is, if the condition (10.11) is no longer satisfied. In this case, exact Coulomb continuum wavefunctions must be used instead of plane waves.

For the calculation of cross sections from the matrix element (10.12) we have to insert for $\varphi_f(\mathbf{r})$ a nonrelativistic Coulomb-distorted wave asymptotically normalized to unit amplitude plus an incoming spherical wave. For a Coulomb potential produced by the charge eZ, the final wavefunction describing an electron emitted with an asymptotic wave number $\mathbf{k}_f$ and direction $\hat{\mathbf{k}}_f = \mathbf{k}_f/k_f$ is given by a partial-wave expansion [140], see also Eq. (3.30).

$$\varphi_f(\mathbf{r}) = 4\pi \sum_{l=0}^{\infty} \sum_{m=-l}^{l} i^l e^{-i\Delta_l} N_{k_f l} (2k_f r)^l e^{-ik_f r}$$

$$\times {}_1F_1(l+1+i\nu, 2l+2; 2ik_f r)\, Y_{lm}(\hat{\mathbf{r}})\, Y_{lm}^*(\hat{\mathbf{k}}_f). \tag{10.19}$$

Here $\hat{\mathbf{r}} = \mathbf{r}/r$ and $\hat{\mathbf{k}} = \mathbf{k}/k$ are unit vectors denoting the angles, $Y_{lm}(\hat{\mathbf{r}})$ are spherical harmonics, and $_1F_1(\cdots)$ is the confluent hypergeometric function. The Coulomb phaseshift is

$$\Delta_l = \arg \Gamma(l+1-i\nu), \tag{10.20}$$

with the Sommerfeld parameter ν given by

$$\nu = \frac{\alpha Z}{k_f \lambdabar_c} = \frac{\alpha Z}{\beta} = \sqrt{\frac{|\epsilon_i|}{\hbar\omega - |\epsilon_i|}} \tag{10.21}$$

and the normalization factor by

$$N_{k_f l} = e^{\frac{\pi}{2}\nu} \frac{|\Gamma(l+1-i\nu)|}{(2l+1)!}. \tag{10.22}$$

When inserting the wavefunctions (10.19) into Eq. (10.12) and adopting the dipole approximation, i.e., replacing $\exp(i\mathbf{k}\cdot\mathbf{r}) \to 1$, only a few partial waves in the expansion (10.19) contribute owing to the dipole selection rules. In this way, Stobbe [141] has derived his well-known formula for the photoelectric cross section per K-shell electron in the form [39]

$$\sigma_{\mathrm{ph}}^{\mathrm{Stobbe}} = \frac{2^8 \pi^2 \alpha}{3} \frac{m_e c^2}{\hbar\omega} \lambdabar_c^2 \left(\frac{\nu^2}{1+\nu^2}\right)^3 \frac{e^{-4\nu \arctan(1/\nu)}}{1 - e^{-2\pi\nu}}. \tag{10.23}$$

The corresponding differential cross section is

$$\frac{d\sigma_{\mathrm{ph}}^{\mathrm{Stobbe}}}{d\Omega} = \sigma_{\mathrm{ph}}^{\mathrm{Stobbe}} \frac{3}{8\pi} \sin^2\theta. \tag{10.24}$$

The Stobbe cross section proves to be quite useful for estimating total cross sections for the atomic photoeffect and for REC up to projectile energies of a few hundred MeV/u, corresponding to electron kinetic energies $(\gamma - 1)m_e c^2$, well below the electron rest energy. This is the case, although the underlying dipole approximation is no longer satisfied and much higher multipoles contribute. By multiplying with the phase-space ratio k'^2/p^2 of emitted photons and emitted electrons, one obtains the cross section for radiative recombination (RR) or radiative electron capture into an empty K-shell $(j_n = \frac{1}{2})$ as

$$\sigma_{\mathrm{RR}}^{\mathrm{Stobbe}} = \frac{2^8 \pi^2 \alpha}{3} \lambdabar_c^2 \left(\frac{\nu^3}{1+\nu^2}\right)^2 \frac{e^{-4\nu \arctan(1/\nu)}}{1 - e^{-2\pi\nu}}, \tag{10.25}$$

where $\lambdabar_c = \hbar/m_e c$ is the Compton wavelength of the electron and $\nu = Ze^2/\hbar v$ is the Sommerfeld parameter (10.21). The constants in front of the ν-dependent terms make up a factor of 9164.7 barn. Cross sections for arbitrary shells have also be calculated, see [135, 140, 142].

10.4   Exact relativistic calculations

The cross sections given in Secs. 10.2 and 10.3 are not valid for high collision
energies and high charge numbers Z, although these approximations for total and sometimes even for differential cross sections carry farther than one
should expect. The exact evaluation of the relativistic photoelectric and
REC cross sections requires a partial-wave expansion of the Coulomb-Dirac
continuum function, see Sec. 6.3. This means that closed-form expressions
can no longer be derived, and one has to resort to numerical methods. Detailed formulations exist since a long time, see e.g., [140, 135]. Since existing
tabulations are not always sufficient in connection with REC, independent
computer codes have been developed and widely applied [143, 144, 137].

The differential cross section for ionizing a single electron in a state n
by an unpolarized photon beam and using electron detectors insensitive to
the spin of the emitted electron is given by [140, 143]

$$\frac{d\sigma^{\mathrm{ph}}(\theta)}{d\Omega} = \frac{\alpha\, m_e c^2}{4\,\hbar\omega}\; \frac{\bar\lambda_c{}^2}{2(2j_n+1)}$$

$$\times \sum_{\mu_n} \sum_{m_s=\pm 1/2} \sum_{\lambda=\pm 1} \left|\, M_{\mathbf{p},n}(m_s,\lambda,\mu_n)\,\right|^2 . \qquad (10.26)$$

Here we have averaged over the $(2j_n+1)$ angular momentum projections μ_n
in the bound state, over the circular polarizations $\lambda = \pm 1$ of the incoming
photon and have summed over the spin components $m_s = \pm\frac{1}{2}$ of the emitted
electron[†]. Because of the summation over all other angular momentum
projections, μ_n, m_s, averaging over the photon polarizations is equivalent
to taking one photon polarization, e.g., $\lambda = 1$ only. However, here we retain
the symmetric formulation involving both values of λ.

In analogy to Eq. (10.12), the relativistic transition matrix element is

$$M_{\mathbf{p},n}(m_s,\lambda,\mu_n) = \int \psi_{\mathbf{p},m_s}^\dagger(\mathbf{r})\, \boldsymbol{\alpha}\cdot\hat{\mathbf{u}}_\lambda\, e^{i\mathbf{k}\cdot\mathbf{r}}\, \psi_{j_n,\mu_n}(\mathbf{r})\, d^3r , \qquad (10.27)$$

where $\hat{\mathbf{u}}_{\pm 1}$ is the unit vector for the circular polarization. The initial bound
state is given by Eq. (6.6) or, specifically for the K-shell, by Eq. (6.18). We
now have two choices of representing the electron continuum wavefunction. One may either quantize the electron spin with respect to the photon
direction or with respect to the electron direction. The former choice, tech-
nically simpler, is applicable only if one sums or averages over the electron
spin. The latter choice, given by the expansion (6.31) is prerequisite if one

[†]In conjunction with an angular momentum representation, it is convenient to sum
over the circular rather than over the linear polarizations $\lambda_{1,2}$.

is interested in the electron polarization. This is so, because a relativistic electron has a sharp spin value only with respect to its own direction of motion. Given the cross section of the photoelectric effect, the cross section for radiative recombination follows from the principle of detailed balance, i.e., by multiplying with the phase-space ratio k'^2/p^2. Since radiative recombination takes place in a moving system, we distinguish its quantities (energy, frequency, angles) by a prime from the unprimed laboratory quantities.

With these definitions, the principle of detailed balance states that the double-differential cross section of radiative recombination is related to that of the photoelectric effect by

$$\frac{\mathrm{d}^2\sigma_{\mathrm{RR}}(E',\theta')}{\mathrm{d}E'\,\mathrm{d}\Omega'} = \left(\frac{\hbar\omega'}{m_ec^2}\right)^2 \frac{1}{\beta^2\gamma^2} \frac{\mathrm{d}^2\sigma_{\mathrm{ph}}(E',\theta')}{\mathrm{d}E'\,\mathrm{d}\Omega'}$$

$$= \frac{(\gamma-1+|\epsilon_n|/m_ec^2)^2}{\gamma^2-1} \frac{\mathrm{d}^2\sigma_{\mathrm{ph}}(E',\theta')}{\mathrm{d}E'\,\mathrm{d}\Omega'}. \qquad (10.28)$$

This equation refers to a specific magnetic substate. If the photoelectric cross section is defined as an average over a subshell, one has to multiply the right-hand side of Eq. (10.28) by a factor $(2j_n+1)$ in order to compensate for this averaging procedure.

In applications to ion-atom collisions, the z-direction is usually defined as the direction of the projectile motion. This is opposite to the direction of the electron momentum as seen from the projectile. Hence for REC, the angle θ' of the photoelectric effect has to replaced by $\pi - \theta'$, or $\cos\theta'$ is replaced by $-\cos\theta'$. In addition to differential and total cross sections, one may also consider details like alignment and polarization, see Sec. 10.6 as well as QED corrections. For a summary, see, e.g., [135].

10.5   REC cross sections from exact calculations

REC compared to radiative recombination

For radiative recombination, that is, for the capture of a free electron with kinetic energy T_e into a specific atomic state with energy ϵ_n, a photon with a well-defined energy for each observation angle, $\hbar\omega = T_e + |\epsilon_n|$ will be emitted. This corresponds to a δ-function in the photon spectrum. However, for REC of a loosely bound electron, the momentum distribution within the target atom represented by the Compton profile (10.8) will lead to a finite line width. Figure 10.4 shows the broadening of the photon spectra for K- and L-REC for two different momentum distributions. The solid line reflects the electron momentum distribution in the nitrogen target

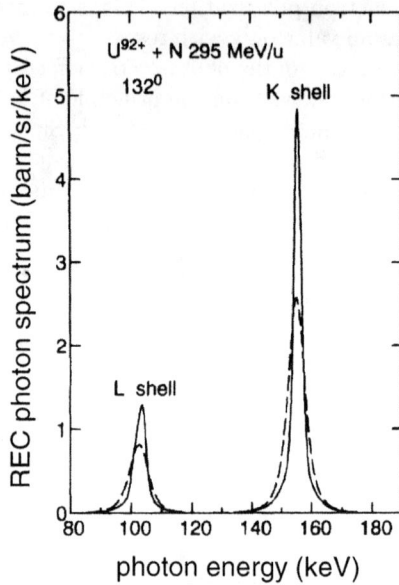

Figure 10.4: Calculated REC photon spectrum for 295 MeV/u U^{92+} on N atoms for the laboratory photon angle of 132^0. Solid line: calculated with approximate Roothaan-Hartree-Fock wavefunctions for the target atom; dashed line: calculated with a hydrogenic momentum distribution (1.21). From [143].

described by Hartree-Fock wavefunctions. Here, the outer electrons are subject only to a screened nuclear charge and hence move more slowly than electrons subject to the full nuclear charge of an unscreened hydrogen-like model (dashed line). As a result, the Hartree-Fock momentum distribution leads to a narrower and sharper photon spectrum.

In order to obtain the single-differential cross section, one has to integrate over the photon line. Since the areas under the photon peaks are approximately the same for delta-function like, hydrogen-like, or Hartree-Fock distributions, the differences between the results from these methods are much smaller than in the photon spectra. Figure 10.5 shows angular distributions calculated, alternatively, from REC (Hartree-Fock) and from radiative recombination. Indeed, for a low-Z target like N_2 or H_2 and a high-Z bare projectile like Au^{79+} or U^{92+}, it turns out that the modification of the cross section due to the electron momentum distribution in

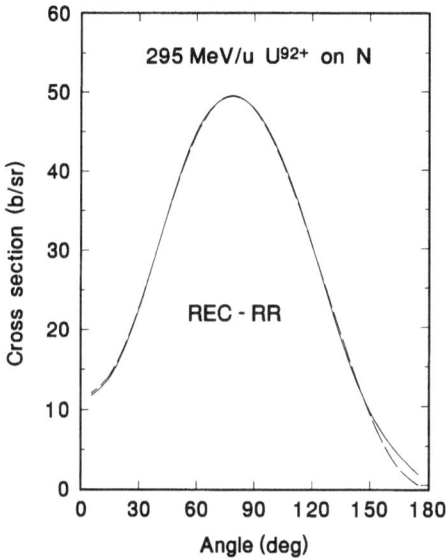

Figure 10.5: Calculated angular distributions for 295 MeV/u U^{92+} on N atoms. Solid line: REC calculated from approximate Roothaan-Hartree-Fock wave functions for the target atom; dashed line: Radiative recombination. Calculations from [145].

the bound target atom is almost negligible, usually of the order of one percent. Therefore, with the present experimental accuracy, it is legitimate to identify REC cross sections with RR cross sections, when comparing to experimental data.

Differential cross sections and effects of the electron spin

Within an exact relativistic formulation, differential cross sections for photoionization cannot be given in a closed form. Therefore, the transformation (10.28) has to be applied numerically to the differential cross sections for photoionization discussed in Sec. 10.4.

Independently of detailed quantitative calculations, one may observe the following features. Quite generally, the differential cross section for radiative recombination into $l = 0$ states of a spinless (nonrelativistic) electron vanishes in the forward and in the backward direction because

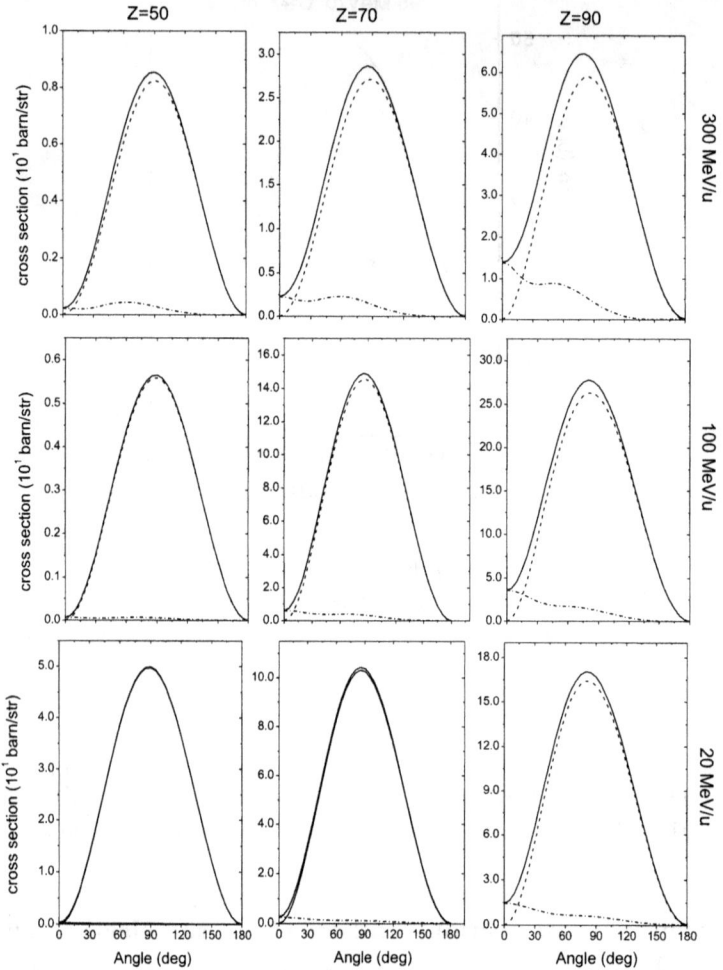

Figure 10.6: Angle-differential REC cross sections for capture into the K-shell of projectiles with charge numbers $Z = 50$, 70, and 90 and projectile energies of 20, 100, and 300 MeV/u. Spin-flip (dot-dash lines) and non-spin-flip (dashed lines) contributions are shown separately. From [146], recalculated in analogy to [137].

initial and final electronic states have $m_l = 0$, so that the emission of a transverse photon with angular momentum ± 1 in its direction of motion is forbidden by angular-momentum conservation. The same is true if the spin is fixed. This means that *radiative recombination and REC into $l = 0$ states at forward or backward angles can occur only by spin-flip processes mediated by magnetic interactions in a relativistic description* [143, 137, 140]. Only spin-flip transitions are able to furnish the angular momentum carried away by the photon. *Finite cross sections at $0°$ and $180°$, therefore, provide a unique signature for spin-flip processes*, which have been predicted to occur for high-Z projectiles and at high collision velocities [143]. Since, except for these spin-flip contributions, the angular distributions with no spin-flip are fixed at the two endpoints to zero, one may expect that deviations from a $\sin^2 \theta$ distribution for K-RR are not very strong even in the moderately relativistic energy range of a few hundred MeV/u.

Experimentally, the effects of spin-flip can be detected in an angular distribution only at forward or backward angles. At an arbitrary angle, one needs other methods of detection, e.g., by capture of polarized electrons into a half-occupied K-shell of a polarized target [147].

In Figure 10.6, we show a set of angular distributions for REC into the K-shell (the deviations of REC for low-Z target atoms from radiative recombination are on the percent level) for three different energies and three different charge numbers [137]. It is seen that with increasing energy and, more importantly, with increasing projectile charge, the deviations from a $\sin^2 \theta$ distribution increase and the spin-flip contributions become more prominent. In the calculations underlying Fig. 10.6 electron partial waves have been included up to $|\kappa| = 20$ and correspondingly high multipole orders in the expansion of the photon field.

In Figure 10.7, we present experimental and theoretical results which exhibit magnetic spin-flip contributions as a clear-cut relativistic effect [148]. In Fig. 10.7(a), the measured differential cross sections for REC into the K-shell of U^{92+} are presented as a function of the laboratory observation angle (solid circles) and compared with predictions based on exact relativistic calculations [143, 137]. In order to achieve convergence, the summation was carried out over all electron partial waves with Dirac quantum numbers $|\kappa| \leq 20$. As seen in the figure, good agreement is obtained between the experimental data and the relativistic theory. In order to elucidate the necessity of a relativistic treatment for high-Z projectiles, the figure also includes the $\sin^2 \theta$ distribution following from a nonrelativistic treatment which incorporates the retardation as well as the Lorentz transformation to the laboratory frame. Obviously, the experimental data deviate considerably from symmetry around $90°$. There are many more experimental

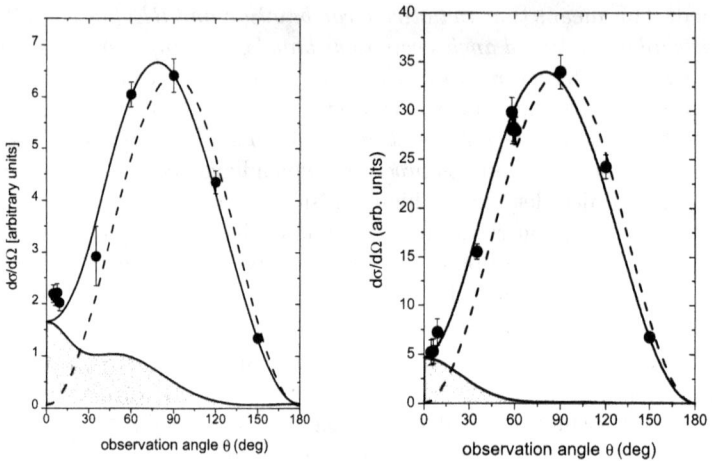

Figure 10.7: (a) Left-hand diagram: Angular distribution for REC into the K shell of bare uranium (solid circles) as a function of the observation angle θ for 309.7 MeV/u U^{92+}+N_2 collisions [148]. The solid line refers to complete relativistic calculations and the shaded area to the spin-flip contributions [146, 143, 137]. The $\sin^2\theta$ shape of the nonrelativistic theory is given by the dashed line. The absolute values of the experimental data and of the nonrelativistic theory are normalized to the result of the complete calculations at $90°$. (b) Right-hand diagram: Angular distributions for K-REC for 88 MeV/u U^{92+}+N_2 collisions [150]. Solid circles: experimental result; solid line: relativistic calculations; shaded area: spin-flip contributions; dashed line: $\sin^2\theta$ distribution.

data for which excellent agreement with the exact relativistic description has been obtained. As an example, we mention measurements for the experimentally resolved $j = 1/2$ and $j = 3/2$ subshells in uranium [149, 150] at low collision energies.

REC as a tool to study the photoelectric effect

From the evaluation of Eq. (10.26) for the photoelectric effect, it follows that at high energies, the angular distribution of the emitted electrons is compressed into a small cone at forward angles, where measurements are difficult. However, if this process is studied as radiative recombination in

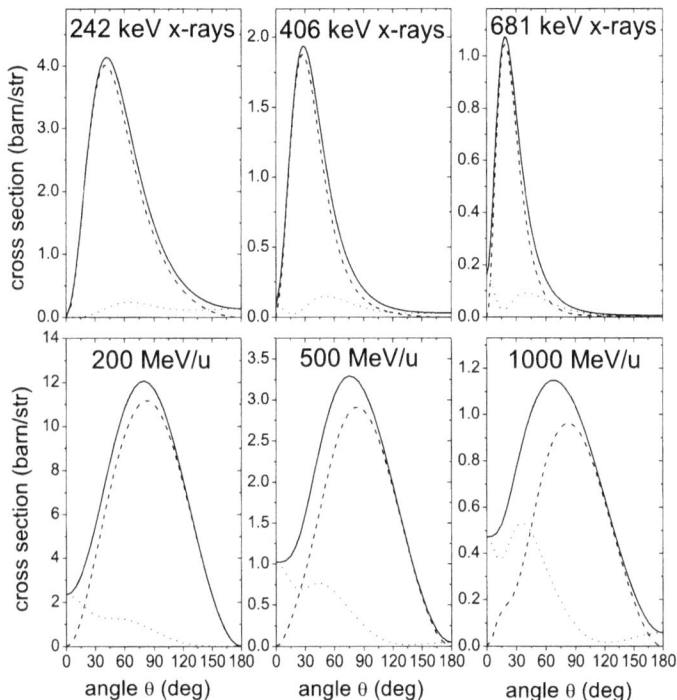

Figure 10.8: Exact angle-differential cross sections for photoionization from the K-shell of hydrogen-like U^{91+} ions (upper row) are compared with the corresponding RR cross sections (lower row). Partial waves up to $|\kappa| = 80$ have been included. The spin-flip (dotted lines) and non-spin-flip (dashed lines) contributions are shown separately. From [146], recalculated in analogy to [151].

the projectile, the transformation (10.17) to the laboratory frame spreads out the distribution over the full angular range, thus rendering it more accessible to measurement. This implies that the incoming photon for the photoelectric effect and the outgoing photon in the corresponding REC reaction have exactly the same energy E_γ.

Examples of such calculations [146, 151] are presented in Figs. 10.8 for photoionization of hydrogen-like U^{91+} and for the corresponding REC process by U^{92+} projectiles, respectively. It is shown that with increasing x-ray energy E_γ, the photon distribution becomes confined to a very narrow cone in the forward direction, which shrinks even further for higher photon energies. The figures also show that at the highest energies, the cross section is dominated by spin-flip transitions. As a remarkable feature, we observe that up to a projectile energy of 1.0 GeV/u, the RR cross sections still roughly follow a sin$^2 \theta$ distribution, the deviations being mainly caused by spin-flip contributions discussed in [137]. This deviation from a simple dipole pattern comes about by the contribution of high multipole components in the expansion of the photon field.

Total cross sections

By integration over the angular distributions, one obtains total cross sections. Here, it is interesting to observe that the nonrelativistic Stobbe cross section (10.25) within the dipole approximation does not depend on Z and v separately, but only on the ratio, that is, on the Sommerfeld parameter ν. In order to exhibit this dependence, it is useful to define an "adiabaticity parameter" η through the relation

$$\eta = 1/\nu^2 \simeq 40.31 \times \frac{E_{\text{kin}}(\text{MeV/u})}{Z^2}. \tag{10.29}$$

The parameter η decides whether a collision is "fast" ($\eta > 1$) or "slow" ($\eta < 1$) and proves to be convenient for presenting RR or REC cross sections per target electron, independently of the collision system. The data plotted in Figs. 10.9(a) and 10.9(b) are compared with theoretical cross-section values. The exact relativistic cross sections for various Z-values do not fully scale according to Eq. (10.29) at higher η-parameters but rather depend on Z and E_{kin} separately. The calculation considers REC into the K, L, and M-shells of the projectile. In general, an excellent agreement between experiment and theory is found in Fig. 10.9(a). The experimental data are, within the error bars, not sensitive to the slight cross section variation predicted by the relativistic theory for different Z systems at one common η-parameter, especially for $\eta \leq 3$. In the higher η-regime, the results of the fully relativistic calculation performed for $Z=80$ deviate already noticeably from the predictions of the nonrelativistic approach. The right-hand figure (b) for high relativistic energies up to about 100 GeV/u shows good agreement with relativistic calculations, while drastic deviations from the nonrelativistic approximation are observed.

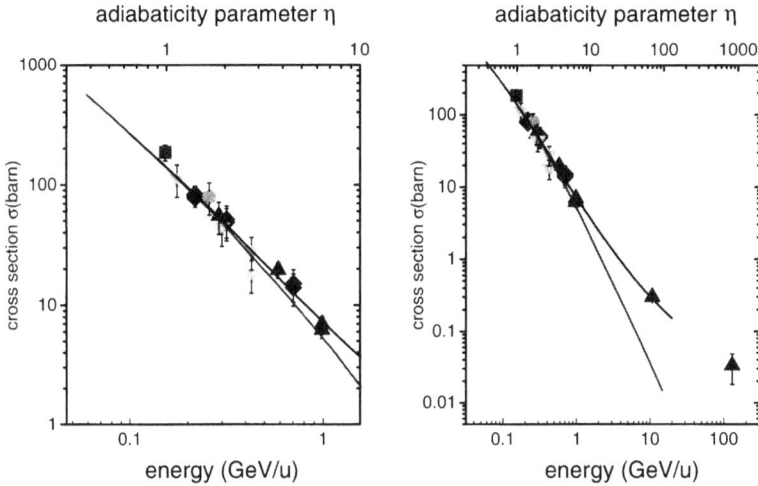

Figure 10.9: (a) Left-hand figure: Total electron-capture cross sections per target electron measured for heavy bare ions ($Z \geq 54$) in collisions with light target atoms (or molecules). The results are plotted as a function of the η-parameter (upper scale) and of the projectile energy (lower scale) and are compared with the result of a relativistic exact calculation for $Z = 80$ (upper line) as well as with the prediction of the non-relativistic dipole-approximation (lower line). (b) Right-hand figure: Same as (a) but extended to high relativistic energies [135].

10.6 Alignment and photon polarization in REC

In Sec. 10.5, we have discussed differential and total cross sections in REC, which implies that the incoming electron in the target atom is unpolarized and the emitted photon is not analyzed with regard to polarization. However, a more detailed picture of the reaction emerges in more sophisticated experiments. For example, we may consider radiative electron capture into an excited state, which subsequently decays by photon emission into a lower state. The angular distribution of the decay photon with respect to the electron direction carries information on the alignment of the intermediate state. In another set of experiments, one may study REC into a given state and measure the polarization of the emitted photon. Again, one will obtain more insight into the reaction. In the following, we discuss both cases.

Alignment

Alignment in an atomic state can be measured by angular correlations when an electron in a continuum state $|\mathbf{p}, m_s\rangle$ is radiatively captured into an excited projectile state $|\kappa_n, \mu_n\rangle$ with the simultaneous emission of an REC photon $|\mathbf{k}, \lambda\rangle$. Subsequently, the intermediate excited projectile state decays into a lower state $|\kappa_i, \mu_i\rangle$, usually the ground state, by emitting another photon $|\mathbf{k}', \lambda'\rangle$. Since the decay photon has a sharp, well-known energy, it can be unambiguously detected experimentally. If the REC photon $|\mathbf{k}, \lambda\rangle$ is not detected and the spin projections as well as the polarization of the decay photon are not observed, on obtains a simple angular correlation between the photon direction $\hat{\mathbf{k}}'$ and the electron direction $\hat{\mathbf{p}}$.

This is a two-step process, which can be best described as the inverse of a two-step photoionization with the angular correlation written as

$$
W(\mathbf{p}, \mathbf{k}') \quad \propto \quad \frac{1}{4} \sum_{\lambda, \lambda' = \pm 1} \sum_{m_s, \mu_i} \int d\Omega_k
$$

$$
\times \left| \sum_{\mu_n} \langle \mathbf{p} m_s | \boldsymbol{\alpha} \cdot \hat{\mathbf{u}}_\lambda e^{i\mathbf{k}\cdot\mathbf{r}} | \kappa_n \mu_n \rangle \langle \kappa_n \mu_n | \boldsymbol{\alpha} \cdot \hat{\mathbf{u}}_{\lambda'} e^{i\mathbf{k}'\cdot\mathbf{r}} | \kappa_i \mu_i' \rangle \right|^2 .
$$

$$(10.30)$$

By introducing density matrices, a multipole decomposition of the photon field and integrating over the unobserved direction $\hat{\mathbf{k}}$ of the REC photon, one derives the explicit angular correlation expressed by anisotropy factors, describing deviations from spherical symmetry [144].

The calculation is usually simplified by assuming that the electromagnetic transition between well-defined atomic states i and n is governed by an electric dipole (E1) matrix element. However, it turned out that experimental and theoretical anisotropy factors indicating an alignment in the intermediate state n, did not fully agree with each other [152]. This discrepancy was resolved when it was discovered that for high nuclear charges Z the magnetic quadrupole (M2) transition may no longer be neglected [153]. Indeed, it was found that for the highest charges, the M2 transition contributes about 30% to the alignment. This is so, because the contribution arises from an interference term between E1 and M2 transitions, while for the total transition probability, the square of the M2 term gives a small contribution compared to the leading square of the E1 term. With the E1–M2 interference correction, good agreement with experimental data is achieved [153].

Photon polarization in REC

We now turn to another way to gain more insight into the dynamics of REC or of RR, namely by measuring the linear or circular polarization of the emitted photon. While alignment measurements are sensitive to substate populations of the intermediate level, measurements of linear polarization are sensitive to interferences between right-hand and left-hand circular polarization of the photon. The basic formalism is very similar [154].

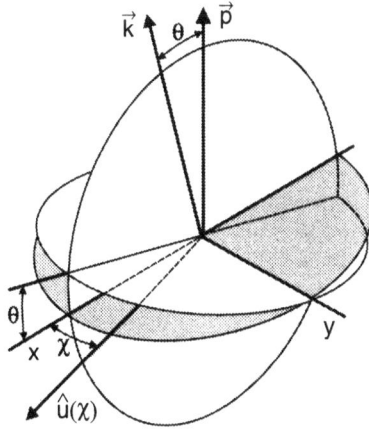

Figure 10.10: Coordinate systems defining the angles θ and χ. The shaded plane is perpendicular to the photon direction and contains the polarization vector $\hat{\mathbf{u}}(\chi)$.

For convenience, we start again with the photoelectric effect in analogy to Sec. 10.4, but discard the summation over λ. The reaction plane is spanned by the electron vector $\mathbf{p}$ and the photon vector $\mathbf{k}$. If the photon polarization is referred to an x-y plane perpendicular to $\mathbf{k}$, see Fig. 10.10, the general angle-dependent cross section for linear polarization in a direction forming an angle χ with the x-axis in the x-y plane is [154]

$$
\begin{aligned}
\sigma_\chi^{\mathrm{ph}}(\theta) &= \frac{\alpha}{4\omega} \frac{1}{2j_n + 1} \sum_{\mu_n} \sum_{m_s = \pm 1/2} \\
&\times \left| \left\langle \mathbf{p} m_s \,\middle|\, \boldsymbol{\alpha} \cdot \tfrac{1}{\sqrt{2}} \left(\mathrm{e}^{-\mathrm{i}\chi} \hat{\mathbf{u}}_+ + \mathrm{e}^{\mathrm{i}\chi} \hat{\mathbf{u}}_- \right) \mathrm{e}^{\mathrm{i}\mathbf{k}\cdot\mathbf{r}} \,\middle|\, \kappa_n \mu_n \right\rangle \right|^2,
\end{aligned}
\tag{10.31}
$$

where $\hat{\mathbf{u}}_\pm = \hat{\mathbf{u}}_{\lambda=\pm 1}$ is the unit vector for the circular polarization $\lambda = \pm 1$.

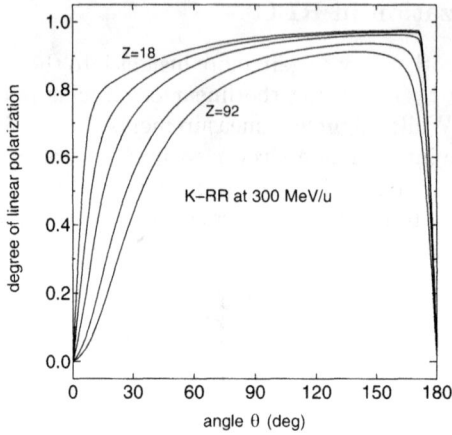

Figure 10.11: Charge dependence of the linear photon polarization in the reaction plane as a function of the emission angle θ for K-RR at a projectile energy of 300 MeV/u. The results are given successively for the charges $Z = 18$, 36, 54, 79, and 92. Calculated from [154].

Specifically, for polarization in the x and y directions, we have $\chi = 0$ and $\chi = \pi/2$, respectively. However, if we want to choose the direction $\hat{\mathbf{p}}$ as z-axis, we have to perform an Euler rotation $\hat{\mathbf{k}} \to \hat{\mathbf{p}}$, mediated by a Wigner rotation matrix, between the old and the new coordinate systems with the angle θ between $\hat{\mathbf{k}}$ and $\hat{\mathbf{p}}$. Correspondingly, the original x-y plane, in which the angle χ is defined, is tilted by the angle θ with respect to the plane perpendicular on $\hat{\mathbf{p}}$, see Fig. 10.10. The azimuth is chosen to be $\varphi = 0$, so that the x-axis lies in the reaction plane.

From the linear superposition of the left-hand and right-hand circular polarizations in Eq. (10.31), we obtain interference terms, which are responsible for the degree of linear polariztion. By introducing a multipole expansion of the photon wave and a partial-wave expansion of the electron continuum function, and subsequently transforming from the projectile frame into the laboratory system, one derives [154] the degree of linear photon polarization in the direction $\hat{\mathbf{u}}(\chi)$ as

$$P_\chi^{\text{lin}}(\theta) = \frac{\sigma_\chi - \sigma_{\chi+\pi/2}}{\sigma_\chi + \sigma_{\chi+\pi/2}} \tag{10.32}$$

Specifically, for linear polarization in the reaction plane, we have $\sigma_\| = \sigma_{\chi=0}$ and for the polarization perpendicular to it, we have $\sigma_\perp = \sigma_{\chi=\pi/2}$.

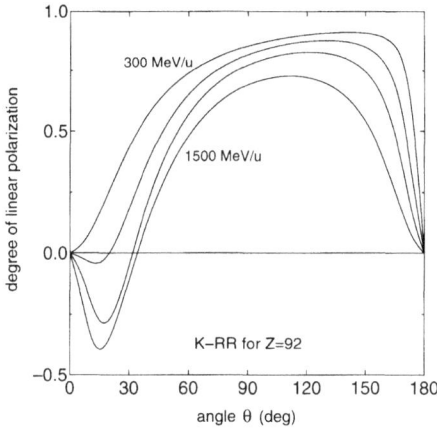

Figure 10.12: Projectile energy dependence of the linear photon polarization in the reaction plane as a function of the emission angle θ for K-RR with $Z = 92$. The results are given successively for projectile energies of 300, 500, 800, and 1500 MeV/u. Calculated from [154].

Figure 10.11 shows the degree of linear photon polarization in the reaction plane ($\chi = 0$) as a function of the emission angle θ for various projectile charge numbers Z and for 300 MeV/u collision energy. One obtains a very high degree of polarization over most of the angular range. The flatness of the polarization for intermediate angles becomes more pronounced as the collision energy decreases. Eventually, in the nonrelativistic limit, the linear polarization in the reaction plane is $P_{\parallel}^{\text{lin}} = 1$, independent of Z. In all calculations, for obtaining a three-digit accuracy, multipole orders up to $L = 17$ are needed and twice as many for the highest energy of 1500 MeV/u in Fig. 10.12

The energy dependence of the photon polarization for a fixed charge $Z = 92$ is illustrated in Fig. 10.12. It is interesting to note that one obtains a "cross-over" at about 500 MeV/u, beyond which the linear polarization becomes increasingly negative at forward angles.

So far, we have considered polarization in the reaction plane, $\chi = 0$. From the explicit results [154], it follows that the χ-dependence is uniquely determined by the angular range from 0° to 45° because

$$P_{\chi}^{\text{lin}} = P_{\pi-\chi}^{\text{lin}} = -P_{\pi/2-\chi}^{\text{lin}} = P_{\chi=0}^{\text{lin}} \cos 2\chi. \qquad (10.33)$$

In particular, the linear polarization vanishes for $\chi = 45°$ and is maximal in the reaction plane.

The degree of circular polarization plays a role only if the incoming electron is polarized and hence carries a definite angular momentum projection, which has to be conserved in photon emission.

Experimentally, the linear polarization of high-energy photons can be detected with high efficiency by a new generation of of segmented germanium detectors, see e.g., [135], allowing for energy as well as for position resolution. In these detectors, advantage is taken of the polarization dependence of the differential cross section for Compton scattering. From the analysis of the scattering data, clear evidence has been derived for a strong polarization of K-REC radiation, in excellent agreement with theoretical predictions.

10.7   Electron capture with pair production

In radiative recombination, an electron is captured into a bound state of a hydrogen-like ion, while at the same time a photon is emitted. However, at high enough energies (in the following, electron energies are always defined to include the electron rest mass) alternative processes may occur [155]. We may imagine that an emitted virtual photon is converted into an electron-positron pair. In other words, if the energy of the incident electron in the nuclear rest frame is larger than the ground-state energy of the corresponding He-like ion plus the positron rest energy, the incident electron can be captured into the $1s_{1/2}$ state with the simultaneous creation of a free-positron $1s_{1/2}$-electron pair:

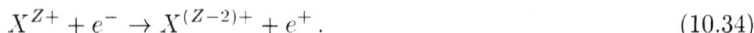

$$X^{Z+} + e^- \rightarrow X^{(Z-2)+} + e^+ . \tag{10.34}$$

This process has also been denoted as "negative-continuum dielectronic recombination" (NCDR) in [155], since it is similar to the usual dielectronic recombination (DR) for a few-electron atom (see, e.g., [156, 157] and references therein). In the latter process, a continuum electron is captured into a target bound state with the energy gain being employed to excite a second electron from the ground state of the target atom to a higher state. This is clearly a resonant process. In negative-continuum dielectronic recombination, the second electron is not an electron already bound to the ion but an electron from the negative continuum ("Dirac sea") which is "lifted" into a bound state. This statement is illustrated in Fig. 10.13. Figure 10.13 (a) indicates the usual DR for a heavy few-electron ion while the NCDR is shown in Fig. 10.13 (b). In this process, we have initially a bare nucleus and an incoming electron, and in the final state we have a He-like ion and an outgoing positron. In contrast to the DR process,

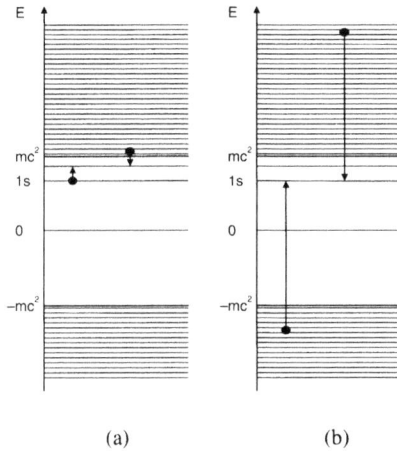

Figure 10.13: Schematic representation of dielectronic recombination (a), of negative-continuum dielectronic recombination into the $(1s_{1/2})^2$ state (b). From [155].

NCDR is not a resonant process owing to the continuum structure of the spectrum at electron energies $\epsilon < -m_e c^2$.

With the increase of the energy (ϵ_i) of the incident electron above the threshold of $\epsilon_i = (\epsilon_{1s_{1/2}})^2 + m_e c^2$, the electrons can occupy excited states as well. When $\epsilon_i > 2 m_e c^2 + \epsilon_{1s_{1/2}}$, the creation of a free-electron–positron pair becomes also possible. We here consider NCDR into the ground state of a He-like heavy ion. For results for this process in the framework of QED see [155].

For the particular case of

$$U^{92+} \ (2.2 \text{ GeV}/\text{u}) + \text{Ne} \rightarrow U^{90+} + \text{Ne}^+ + e^+ \qquad (10.35)$$

the cross section per target electron is $\sigma \approx 2.7 \times 10^{-5} \text{ b} \approx 8.4 \times 10^{-6} \ \sigma_{\text{REC}}$. Although this cross section is exceedingly small, the signature of charge change by two units and positron emission renders the new process clearly distinct from all other types of electron capture (e.g. radiative recombination), and it should become observable with the advent of the next generation of heavy-ion storage rings with higher energy and a much higher beam current, as planned, for example, at the future Facility for Antiproton and Ion Research (FLAIR) at the GSI Darmstadt.

Chapter 11

Electron-positron pair production

The production of electron-positron pairs in collisions of charged cosmic ray particles with nuclei has been a subject of experimental and theoretical interest soon after the discovery of the positron by C.D. Anderson in 1932. In an intuitively appealing picture, the equivalent-photon or Weizsäcker-Williams method, see, e.g., [138, 97, 3], we may imagine that the rapidly changing electromagnetic field of the moving particle with respect to the laboratory frame is Fourier-decomposed into pulses of polarized radiation, see Sec. 5.6. The projectile is hence replaced by a swarm of high-energy, almost real photons, i.e. with $kc \approx \omega$, where k is the wave number and ω the frequency. If b is the impact parameter, the approximate duration of the passage is given by $\Delta t \approx b/\gamma v$, so that the photon spectrum extends up to a maximum frequency $\omega_{\max} = \gamma c/b$ given in Eq. (5.40), which increases dramatically for high values of the Lorentz factor γ, see Table 5.1. A simple estimate [138] for the total pair production cross section at large values of γ is provided by

$$\sigma_{e^+e^-} = \frac{28}{27\pi}(\alpha Z_{\mathrm{T}})^2(\alpha Z_{\mathrm{P}})^2 \lambda_{\mathrm{c}}^2 \left(\ln \frac{\gamma}{4}\right)^3, \tag{11.1}$$

where λ_c is the electron Compton wavelength and α is the fine-structure constant. The cross section increases monotonically with γ and becomes large for high-Z collision partners.

In the near future, very high collision energies in the center-of-mass system can be attained in colliders, in which counter-propagating beams of like particles with equal and opposite velocities collide. For a given collider Lorentz factor γ_{coll}, the equivalent projectile Lorentz factor γ_{FT} for a fixed

Table 11.1: Estimates from Eq. (11.1) for pair production at an impact parameter $b = 386$ fm equal to the electron Compton wave length. For fixed-target (FT) accelerators, the laboratory energy is given, for the heavy-ion colliders RHIC (Brookhaven) and LHC (Geneva), typical projectile energies E in the collider frame are listed. The maximum photon energies $\hbar\omega_{max}$, (5.40) and Table 5.1, are presented together with the estimated pair-production cross section σ_{pair} according to Eq. (11.2) [158].

	E	$\hbar\omega_{max}$ at λ_c	nuclei	σ_{pair}
	(GeV/u)	(MeV)		(kb)
FT	5	3.3	U – U	0.01
FT	10	6.0	U – U	0.12
FT	30	17.0	U – U	0.95
RHIC	100	60	Au–Au	30
LHC	3 400	2 000	Ca- Ca	0.7
LHC	3 400	2 000	Pb–Pb	200

target is given by Eq. (5.16) as $\gamma_{FT} = 2\gamma_{coll}^2 - 1$. In this way, extreme relativistic collision energies can be reached, for which pair production is one of the dominating processes. Estimates for pair-production cross sections are presented in Table 11.1.

The pair production process can be best visualized by considering the spectrum of the Coulomb-Dirac equation, see Fig. 11.1. The creation of a pair corresponds to the excitation of an electron from the negative-energy sea to a bound or a continuum state of positive energy. The former case is denoted as "bound-free pair production", the latter as "free pair production". We start in Sec. 11.1 with discussing the production of free electron-positron pairs in a QED description and subsequently in a perturbative approach that takes into account the Coulomb distortion of the electron- and

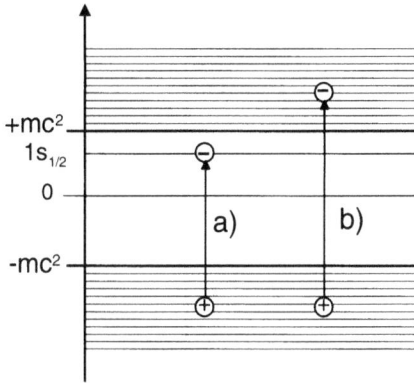

Figure 11.1: Schematic representation of (a) bound-free pair production and (b) free pair production.

positron wavefunctions. We then proceed in Sec. 11.2 to bound-free pair production in perturbative and non-perturbative approaches. In particular, we point out the difficulties associated with describing a negative-energy time-dependent two-center continuum. Finally, in Sec. 11.3, we address multiple pair production at extremely high collision energies.

11.1 Production of free electron-positron pairs

QED- and equivalent-photon description

Energy-momentum conservation forbids the creation of an electron-positron pair by a single moving charge P. Therefore, a second particle T is needed to take up the excess energy and momentum. The lowest-order QED process can be described as follows: Nucleus P or T emits a virtual photon, which disintegrates into an e^+e^- pair. One of the leptons leaves the collision region freely without further interaction, while the other lepton interacts with the other nucleus, T or P, respectively, by exchanging a virtual photon (carrying energy and momentum) and subsequently leaves the interaction region freely. An illustration in terms of Feynman diagrams is given in Fig. 11.2.

The first exact QED description based on the diagrams of Fig. 11.2 has been worked out by Bottcher and Strayer [159]. This treatment is adequate for high collision energies and hence high lepton momenta, when the leptons

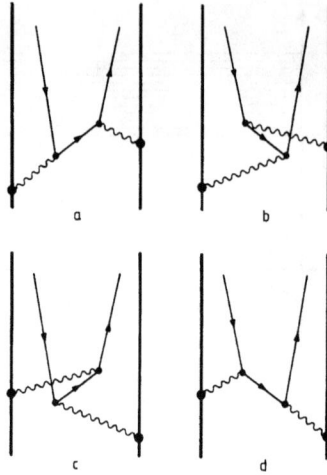

Figure 11.2: Lowest-order Feynman diagrams for the production of a free electron-positron pair in the collision between two heavy ions indicated by heavy lines. The photons are indicated as wavy lines, while the electrons and positrons are described by thin lines with the arrows directed upwards and downwards, respectively.

can be described as undistorted plane waves. It requires the evaluation of 9-dimensional integrals by Monte-Carlo techniques. The path of employing Feynman integral techniques in QED calculations has been further followed in [160, 161, 162].

In earlier calculations, the equivalent-photon approximation has been employed, see Sec. 5.6. Including corrections of the pulse P_2 in addition to the leading term P_1, Racah (1937) [163, 158], has derived the lowest-order total cross section in an analytical form as

$$\sigma = \frac{Z^4 \alpha^4}{\pi m_e^2} \frac{28}{27} \left(\ln^3 \gamma_{coll}^2 - 2.19 \ln^2 \gamma_{coll}^2 + \cdots \right), \tag{11.2}$$

where $Z_P = Z_T = Z$ and a collider kinematics has been assumed. These results are in close agreement with those of present-day Monte-Carlo calculations [159] and – for higher energies – with recent numerical evaluations of the equivalent-photon method, see [164].

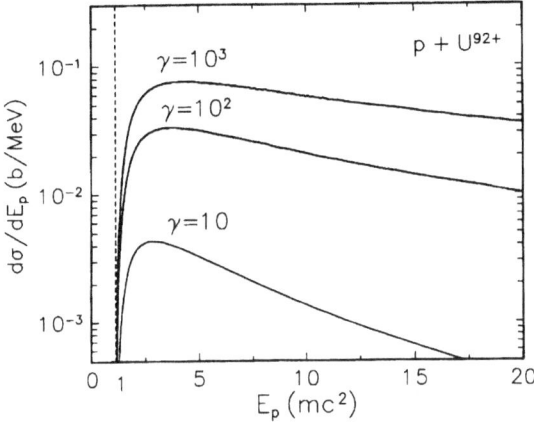

Figure 11.3: Single-differential cross sections for P + U^{92+} collisions as a function of the positron total energy E_p for several values of the Lorentz factor γ. For other projectiles, the perturbative cross section scales as Z_P^2. From [167].

Perturbative treatment with Coulomb distortion

For lower relativistic energies, say up to about 10 GeV/u, and high-Z nuclei, one has to take into account the Coulomb distortion of the lepton waves by the nuclei, adopting a distorted-wave Born approximation (DWBA) [165, 166, 167]. In line with Fig. 11.1, the transition amplitude (in natural units) is given by Eq. (7.9) and can be rewritten as

$$A_{p_+p_-}(\mathbf{b}) = i\frac{\gamma Z_P e^2}{\hbar} \int dt\, e^{i(E_{p_+}+E_{p_-})t/\hbar}$$
$$\times \int d^3r\, \varphi_{\mathbf{p}_+}^\dagger(\mathbf{r})(1-\beta\alpha_z)\frac{1}{r'}\varphi_{\mathbf{p}_-}(\mathbf{r}), \qquad (11.3)$$

where E_{p_+}, $\mathbf{p}_+$ and E_{p_-}, $\mathbf{p}_-$ are the electron and positron energies and momenta. The sixfold-differential pair production cross section can be stated in the form

$$\frac{d^6\sigma}{dE_{p_+}\,dE_{p_-}\,d\Omega_{p_+}\,d\Omega_{p_-}} = \frac{p_+p_-E_{p_+}E_{p_-}}{c^4}\sum_{s_+s_-}\int d^2b\,|A_{p_+p_-}(\mathbf{b})|^2,$$

$$(11.4)$$

with the summation extending over the unmeasured spin projections of electron and positron. The evaluation proceeds similarly as in Sec. 4.1:

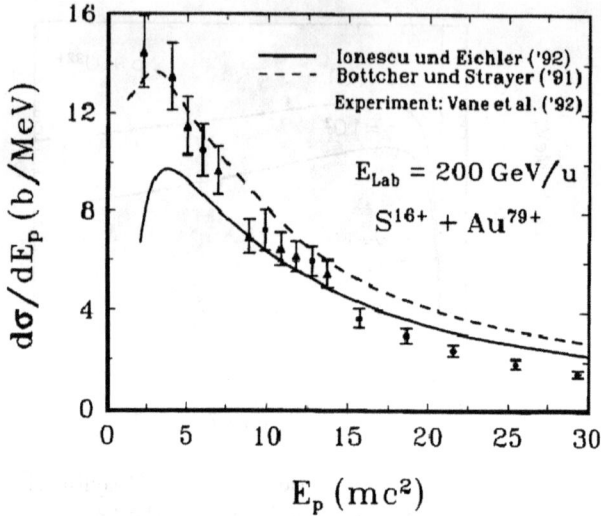

Figure 11.4: Single-differential cross sections in S^{16+} + Au^{79+} collisions at $E_{\text{lab}} = 200$ GeV/u as a function of the total positron energy E_p. Dashed curve: QED calcualtions [159] (not including final-state interactions); solid curve: first-order perturbation theory with Sommerfeld-Maue wavefunctions including final-state interactions [167]. Experimental values are from [168].

One uses appropriate Fourier transforms and the generalized Bethe integral (8.2) to move the space-time dependence in $A_{p_+p_-}(\mathbf{b})$ to the exponent and then performs the space-time integrations.

Compared to the case of excitation or ionization, the evaluation of the amplitudes (11.3) is complicated by the presence of two continuum Coulomb-Dirac wavefunctions. By integrating over the directions of emission of positron and electron, one obtains the double-differential cross section $\mathrm{d}^2\sigma/\mathrm{d}E_{p_+}\,\mathrm{d}E_{p_-}$ depending only on the electron and positron energies. If a partial-wave expansion of the exact continuum states is used, one obtains an *incoherent* summation over partial waves [165] for the double-differential cross section. In practice, owing to computer limitations, the summation over partial waves has to be truncated at a certain maximum angular momentum for electron and positron, thus limiting the maximum projectile energy to which this method can be applied.

The partial-wave expansion can be avoided by using approximate Furry-

Sommerfeld-Maue wavefunctions (6.32) [166, 167], valid for lepton energies large compared to the rest energy $m_e c^2$. In Fig. 11.3, we show the behavior of the single-differential cross section obtained by integrating over electron

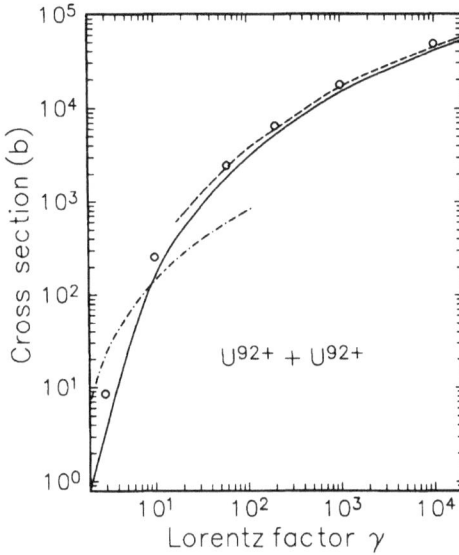

Figure 11.5: Total cross sections for pair production as a function of the (fixed-target-) Lorentz factor γ in collisions between bare uranium nuclei. Dash-dotted curve: first-order perturbation theory with partial-wave expansion [165]. Dashed curve: QED calculations [159]. Circles: combination of perturbative QED calculations with the equivalent-photon method [169]. Solid curve: first-order perturbation theory with Sommerfeld-Maue wavefunctions. From [167].

energies [167]. It is seen that the positron spectra extend to increasingly high energies as the projectile energy increases. Figure 11.4 gives a comparison between theoretical and experimental single-differential cross sections $d\sigma/dE_p$ for the collision system $S^{16+} + Au^{79+}$ at a laboratory energy of 200 GeV/u as a function of the total positron energy. The different symbols represent measured cross sections at different magnetic field settings. It is seen that the theories approximately reproduce the experimental behavior, with the QED calculations being better at lower and the distorted-wave calculations better at higher energy. In view of experimental uncertainties, it is, however, premature to distiguish between the theoretical approaches.

The total cross section is obtained by integrating also over the positron spectrum. Typical theoretical results for $U^{92+} + U^{92+}$ collisions are presented in Fig. 11.5. One notices that the partial-wave expansion fails at higher energies, because computer limitations set a limit to the highest partial wave to be included. At very high energies, the results of QED and of calculations including final-state interactions are very close to each other, as to be expected.

11.2   Bound-free pair production

Perturbative treatments

In a pair production process, the electron may be created in a *bound state* of the target or the projectile, rather than in a continuum state, see Fig. 11.1. This requires that its momentum can be accommodated in the bound-state momentum distribution. Since K-shell wavefunctions offer the broadest momentum spread, production of the electron in the K-shell is dominant. Experimentally, this new process is detected by the decrease of the original projectile charge state by one unit [170, 171, 172] and is sometimes interpreted as a capture process, see [173].

In contrast to free pair production, for which the asymptotic dependence on the projectile energy and on the charge states involved is given by Eq. (11.1), the asymptotic dependence of the cross section for pair production with the electron emerging as bound to the projectile is given by [121]

$$\sigma_{\mathrm{fi}}^{\mathrm{b-f\ pair}} \propto Z_{\mathrm{T}}^2 Z_{\mathrm{P}}^5 \ln\left(\frac{\gamma}{\gamma_0}\right), \tag{11.5}$$

where γ_0 varies between 6.81 and 8.23 as the projectile charge increases from $Z_{\mathrm{P}} = 1$ to $Z_{\mathrm{P}} = 92$. The factor Z_{T}^2 arises from the perturbative action of the target nucleus while the factor Z_{P}^5 reflects the availability of high-momentum components in the bound-state wavefunction of the projectile and similarly occurs for charge transfer, see Eqs. (9.3), (9.10), and (9.12).

Bound-free pair production is of particular relevance for the design and operation of colliders, because, even in an ideal vacuum, it changes the charge state of the colliding ions, which subsequently get lost from the beam in the storage ring, thus limiting the beam's luminosity and lifetime. The process has been extensively treated with perturbative and nonperturbative methods. As perturbative calculations, we mention [97, 174, 175, 176]. Baltz et al. [176] predict the total cross section for bound-electron – free-positron pair production to be of the form

$$\sigma_{\text{fi}}^{\text{b-f pair}} = A \ln \gamma + B, \tag{11.6}$$

where $\gamma = \gamma_{\text{FT}}$ is the equivalent fixed-target Lorentz factor (5.16) and the coefficients A and B are independent of γ (to within higher orders of $1/\gamma^2$). The quantity A has to be estimated from perturbation theory, while B receives contributions both from the perturbative and the nonperturbative range of impact parameters. Note that the cross section derived from a perturbative description of pair production has a similar behavior as for ionization, see Eq. (8.8).

The cross section for capture of the electron into different bound states has been investigated as a function of the charge state for the high energies of the forthcoming colliders [177]. The calculations confirm the $1/n^3$ scaling law for capture into ns states.

Nonperturbative treatments

Since for bound-free pair production only one of the leptons is in the continuum, this case is more easily amenable to nonperturbative treatments. Coupled-channel calculations require discrete basis states. If continuum states are involved, one, therefore, has to discretize the continuum, for example by introducing "pseudostates", see Sec. 2.10. These states are represented by arbitrarily chosen square-integrable functions selected so as to improve the flexibility of the basis set in approximating the true two-center wavefunction throughout the collision. They are not eigenstates of the unperturbed Hamiltonian. Another method, which allows one to understand better which part of the continuum is included, is provided by wave packets constructed as superpositions of continuous eigenstates.

In constructing these wave packets, see Sec. 7.3, one first segments a certain energy range of the continuum into finite intervals, centered around the energy E_k and with the width ΔE_k. If one superimposes time-dependent wavefunctions, $\psi_{E_k}(\mathbf{r}_{\text{T}}, t) = \varphi_{E_k} \exp(-iE_k t)$, one obtains Weyl wave packets, see e.g., [178, 3]. In actual calculations, however, one usually adopts stationary wave packets (7.11)

$$\varphi_{E_k}(\mathbf{r}_{\text{T}}) = \frac{1}{\sqrt{\Delta E_k}} \int_{E_k - \Delta E_k/2}^{E_k + \Delta E_k/2} \varphi_E(\mathbf{r}_{\text{T}}) \, dE. \tag{11.7}$$

Owing to the superposition of adjacent continuum functions, the fall-off of
the radial wavefunctions is much faster than the decrease $1/r$ of the functions $g(r)$ and $f(r)$ themselves. An example of a wave packet is illustrated
in Fig. 7.3.

Single-center coupled-channel calculations have been performed in [179,
180, 181], finite-difference calculation for a direct numerical solution of the
time-dependent Dirac equation in [111], see Sec. 7.6, and lattice calculations
in [182, 112]. Similarly, the Dirac equation in *momentum space* has been
solved numerically in [113, 115], see Sec. 7.7. Another treatment has been
based on the Magnus approximation to the time-evolution operator, see
Sec. 3.2, [183].

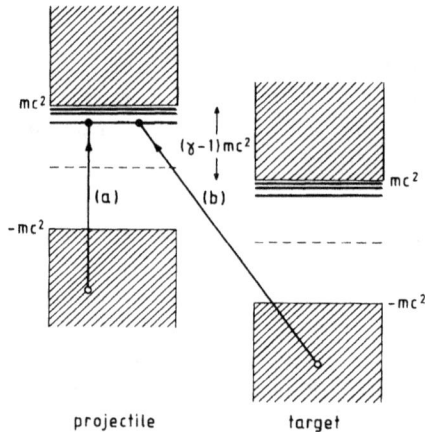

Figure 11.6: Schematic diagram illustrating two mechanisms for bound-free pair
production: (a) excitation type and (b) transfer type simulating the two-center
continuum [184].

The negative-energy two-center continuum

Almost all coupled-channel calculations for pair production have been performed with single-center basis sets, which allow one to include a very large
number of basis states [179, 180, 181]. However, while in single-center calculations one of the ions is considered to generate the eigenstates and the
other ion just provides the perturbation, a general description must treat
target and projectile on equal footing. This can be achieved by two-center
coupled-channel calculations, e.g., [109, 134]. There are two difficulties

with these calculations: (a) Because of time-dependent overlap and interaction matrix elements, subject to Lorentz contraction, it has not been
possible so far, to reach convergence for pair production calculations. (b)
It is not easily possible at present to describe a time-dependent two-center
continuum, which can only be simulated by attaching a continuum to each
of the centers. For bound-free pair production, one therefore can distinguish two contributions illustrated in Fig. 11.6, namely excitation-type pair
production in which a hole is created in the negative-energy sea of the target, and transfer-type pair production, in which the hole is created in the
negative-energy sea of the projectile. This mechanism has been proposed
by Eichler [184] and further worked out in [185] within a perturbative description, which treats the two contributions independently.

However, the only proper way to treat excitation-type and transfer-type
pair production in a unified manner, is by two-center calculations. It is then
possible to separate both contributions [134]. Figure 11.7 shows the result

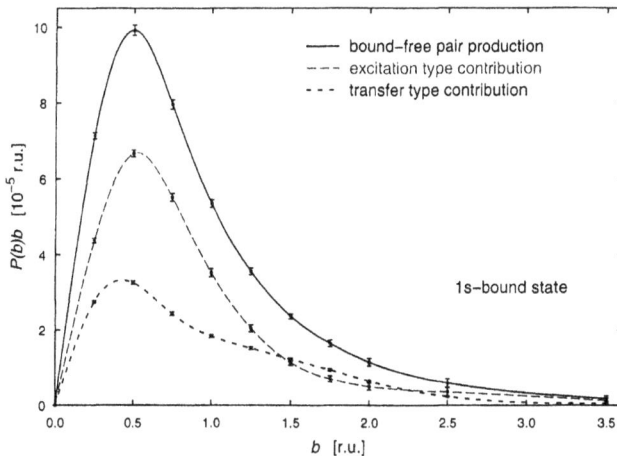

Figure 11.7: Weighted probabilities for bound-free pair production in U^{92+} +
U^{92+} collisions at 1.0 GeV/u as a function of the impact parameter obtained for
coupled-channel calculations in the collider frame using a two-center basis of 68
basis functions [134].

of two-center coupled-channel calculations in the collider frame for bound-
electron free-positron pair creation as a function of the impact parameter
for collisons of bare uranium nuclei at 1.0 GeV/u. In this case, in addition

to 10 bound-state basis functions, 24 wave-packet basis functions (11.7) for each center have been included with $\kappa = \pm 1$, $E_k = \pm 1.15$, ± 1.45, ± 1.75 and $\Delta E_k = 0.3$ (all in relativistic units, r.u.), half of them with positive energy and half with negative energy. The radial wavefunctions of these wave packets are approximately localized within a sphere of 200 r.u. By comparison with Fig. 9.6, it is realized that the maximum of the probability occurs at smaller impact parameters than for charge transfer. This is an obvious consequence of the fact that pair production demands high field strength, higher than in the order-of-magnitude estimates of Table 11.1. One also notes that the energy of the wave packets included in the basis set, is not sufficient to calculate a positron spectrum. Insofar, the results are of a qualitative nature.

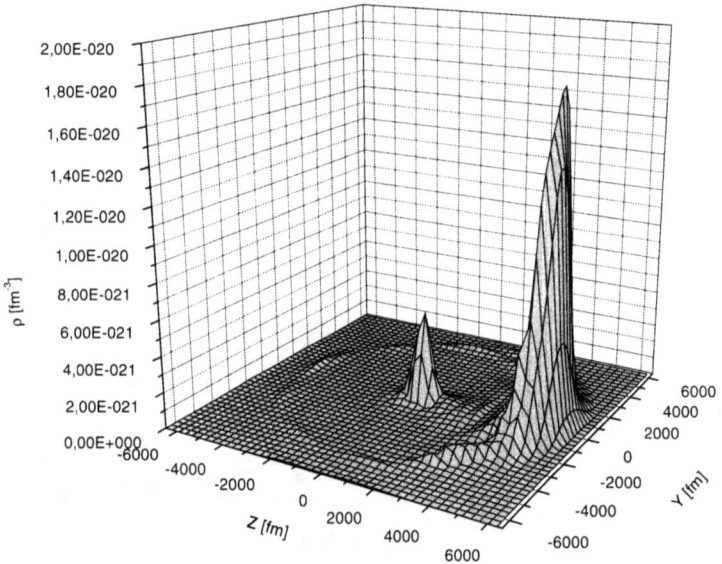

Figure 11.8: Probability density of the negative-energy continuum states for the collision Au^{79+} on U^{92+} at $\gamma = 2$ and $b = 530$ fm. The projection has been performed for quantum numbers $|\kappa| \leq 5$. From [112].

Most importantly, it is seen that the transfer-type pair production provides a significant contribution to the total bound-free pair production. Although the computations go to the limit of the available computer power, convergence is certainly not achieved. Nevertheless, the results clearly demonstrate the importance of a molecular (two-center) continuum for the negative-energy states. This is analogous to the representation of molecular states in atomic-orbital expansions as discussed in Sec. 2.9 for nonrelativistic collisions.

The difficulty in treating negative-energy states in time-dependent two-center system can also be traced in the nonperturbative calculations on a lattice by Busić. et al. [112]. By construction, a lattice is centered about one of the collision partners, say the target. Therefore, positive- and negative-energy components are referred to the target coordinate system, thus giving up the symmetry between both partners. Figure 11.8 shows the probability density of the negative-energy continuum states for the collision Au^{79+} on U^{92+} at $\gamma = 2$ and $b = 530$ fm. A prominent peak is seen near the position of the receding projectile, which might suggest the emissions of positrons. However, as a closer analysis proves [112], this conjecture is misleading: a large part of the negative-energy contributions stems from *bound* projectile states populated by charge transfer but giving rise to sizeable negative-energy components in the *target frame*. This again points to the difficulty in describing negative-energy states in a two-center system.

Coulomb boundary conditions and extreme-relativistic collisions

A great deal of attention has been devoted to modifying the original Dirac equation in such a way that the effective interaction becomes short-range and one hence achieves better convergence in coupled-channel calculations [109] or in direct numerical solutions of the time-dependent Dirac equation [112]. The method applied in various treatments is essentially the same but has been given different names. It has been denoted as an approach satisfying "Coulomb boundary conditions" by introducing a suitable phase transformation (for a discussion see Sec. 7.4 and the review [49]) and has been first analyzed in coupled-channel calculations for relativistic collisions in [109]. The method can also be interpreted as a "gauge transformation" [176, 186, 187, 188, 189, 190, 191] formulated to render the problem more tractable. In [190], an exact numerical solution of the time-dependent Dirac equation for asymptotically high collision energies was obtained and confirmed independently in lattice calculations [112]. In [191] a new light-fronts approach for treating nonperturbative effects at asymptotically high energies was derived. It has been pointed out in connection with coupled-channel calculations [192] that a gauge transformation cannot change the physical content or parametric dependencies [187], but may improve the convergence properties of basis expansions. A detailed analysis of phase transformations of this kind is given in [193].

11.3 Multiple pair production

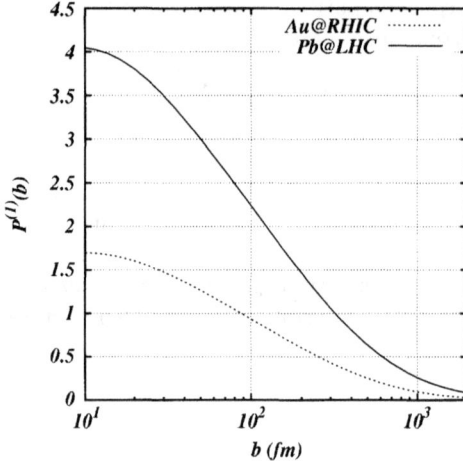

Figure 11.9: Violation of unitarity at small impact parameters for Au + Au at RHIC energies ($\gamma_{\mathrm{FT}} = 2 \cdot 10^4$) and Pb + Pb at LHC energies ($\gamma_{\mathrm{FT}} = 3 \cdot 10^7$). From [158].

Limitations of perturbative calculations became evident already in early work of Bethe, Maximon, and Davies [104, 194] taking into account the final-state interaction of the Coulomb field. Disregarding Coulomb contributions proportional to the logarithm $\ln \gamma$ of the Lorentz factor and assuming $E_e, E_p \gg m_e$, one obtains an approximate analytical expression [195] for the first-order impact-parameter dependent pair production probability in the form

$$P^{(1)}(b) = \frac{14}{9\pi^2} (\alpha Z_{\mathrm{T}})^2 (\alpha Z_{\mathrm{P}})^2 \frac{\lambda_c^2}{b^2} \ln^2 \left(\frac{\gamma d \lambda_c}{2b} \right), \qquad (11.8)$$

where $d = 0.681$. This expression is valid for $\lambda_c \leq b \leq \gamma d \lambda_c$. For Pb + Pb collisions at $\gamma_{\mathrm{FT}} = 3 \times 10^7$, see Eq. (5.16), and $b = \lambda_c$, one obtains an unrealistic probability $P^{(1)}(\lambda_c) = 5.3$, which clearly shows the need for a nonperturbative treatment. More accurate calculations [158] yield an impact-parameter dependent transition probability that clearly shows the violation of unitarity at small impact parameters, see Fig. 11.9.

For highly-charged collision systems as for example $U^{92+} + U^{92+}$ and extremely high collision energies, one therefore expects that creation of multiple pairs in a single collision will occur. Indeed, it has been shown [195] that the production of multiple pairs in a single collision restores unitarity. There are several theoretical treatments of this process [195, 196, 197, 198, 199]. While these approaches are quite different in detail, they all adopt

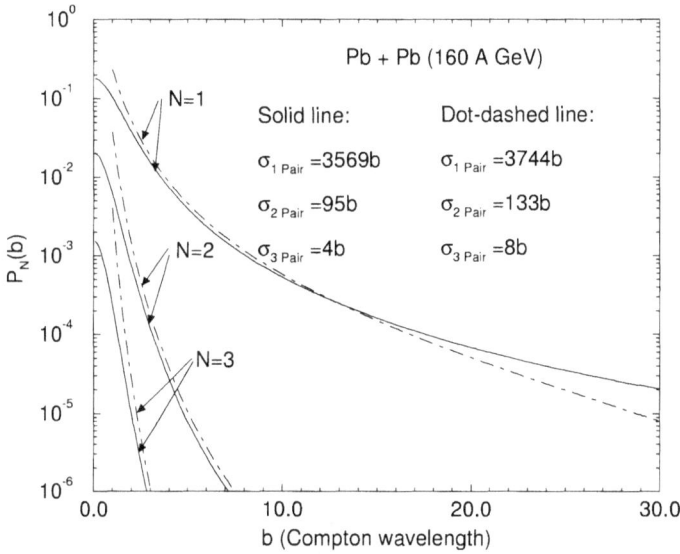

Figure 11.10: Impact-parameter dependent probability distribution for N-pair production in Pb + Pb collisions at 160 GeV/u in the fixed-target frame. Solid lines: exact numerical results, dash-dot lines: equivalent photon approximation. Note that the K-shell radius $a_K(\text{Pb})$ is approximately 1.7 times the Compton wave length, see Eq. (1.3). From [164].

the basic assumption, without which a theory would become exceedingly complicated, namely that the pairs are created completely independently of each other. This means that electrons and positrons do not interact with one another (this would be a $1/Z$ correction) and are not subject to the Pauli principle. The resulting multipair production probability is described

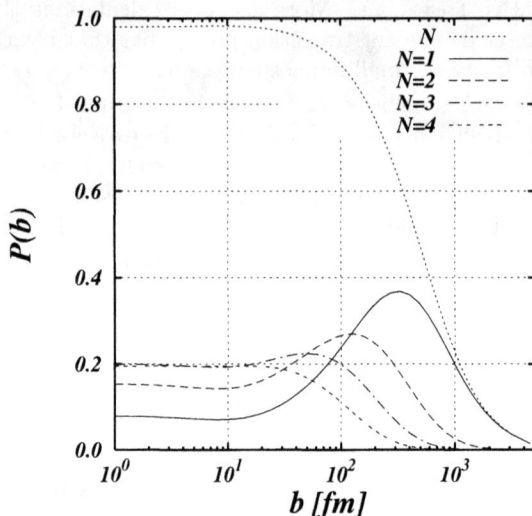

Figure 11.11: Impact-parameter dependent probability distribution for N-pair production in Pb + Pb collisions at $\gamma_{\text{lab}} = 2950$. Also shown is the probability $\sum_N P(N, b) = 1 - P(0, b)$ to produce at least on $e^+ e^-$ pair. From [98].

by a Poisson distribution

$$P(N, b) = \frac{[P^{(1)}(b)]^N}{N!} \, e^{-P^{(1)}(b)}, \tag{11.9}$$

where N denotes N-pair production and $P^{(1)}(b)$ is the impact-parameter dependent lowest-order perturbation result for the total single-pair production probability. This quantity can also be interpreted as the average number of electrons produced out of the vacuum.

In Fig. 11.10, we present the N-pair probability distributions [164] calculated from exact evaluations of a QED theory. Also shown are results from the equivalent-photon approximation, see Sec. 5.6, according to [198]. It can be observed that single-pair creation dominates the picture. The N-pair production probabilities with $N \geq 2$ are seen to decrease very rapidly for the depicted impact-parameter range $b \geq \lambda_c$.

It is instructive to consider the probability distributions at very small impact parameters, where Eq. (11.8) is no longer valid. Applying an approach applicable to impact parameters as small as $b \approx 2R$, corresponding to the case that the nuclei touch each other, Baur et al. [98] obtain the probability distribution displayed in Fig. 11.11. It is noted that for nearly grazing collisions, on the average 3–4 pairs will be produced. Owing to their low energy and their low weight $P(N,b)b$, they will usually escape observation. It follows that the total cross section is dominated by larger impact parameters and will not be affected by the contributions at small b.

In any case, the experimental detection of multiple pair production will be very difficult, and so far no evidence has been found.

Among all the atomic reactions occurring in energetic collisions, electron-positron pair production and multiple pair production are the only processes that can occur to an appreciable extent only at relativistic ion velocities.

Part III

Selected topics

Chapter 12

Selected Topics I: Hyperspherical coordinates

In the current and in the following Chapters, I wish to include a few subjects that are outside a systematic treatment of ion-atom collisions but rather represent extensions and applications. The individual Chapters are not connected to each other and are designed to give a simple first insight into fields or methods that have found wide interest in recent years.

In the preceding Chapters, we have mostly considered a system composed of two nuclei and a single active electron. By treating the nuclear motion classically, it is possible to reduce the dynamics of the system to that of a time-dependent single-electron problem. This approach is justified in many situations, in particular for inner-shell problems. However, there are, of course, circumstances in which more electrons are involved and in which electron correlations play a role. A powerful method to cope with many-electron systems including correlations is based on a description within "hyperspherical coordinates" the essence of which is discussed in the current Chapter.

In Sec. 12.1, we discuss in some detail the simplest case, namely the system composed of a nucleus and two correlated electrons, then proceed in Sec. 12.2 to the adiabatic approximation in the form of a Born-Oppenheimer separation and discuss its application within the close-coupling method in Sec. 12.3.

12.1   Hyperspherical coordinates for two-electron systems

In the present Lectures, I have explicitly treated collisional three-body systems, but I have not incorporated interactions between *all* three particles. Indeed, the assumption of straight-line trajectories ignores the ion-ion interaction. While this is well justified for energetic collisions, there are problems for which all interactions have to be taken into account.

Quite generally, one may deal with larger aggregates of electrons and nuclei. In such a case, each configuration of N particles may be represented by a single vector $\mathbf{R}$, which stands for $3(N-1)$ internal coordinates referred to the center of mass. The important quantity is the modulus R of $\mathbf{R}$, which represents the aggregate's overall size, while the multidimensional direction $\hat{\mathbf{R}}$ specifies its internal geometry. In particular, the single global parameter R, denoted as "hyperradius", controls the evolution of the aggregate from a compact to a fragmented structure. In general, the hyperradius of an N-body system composed of masses $m_1, m_2, \cdots, m_N$ located at the coordinates $\mathbf{r}_1, \mathbf{r}_2, \cdots, \mathbf{r}_N$ with respect to the center of mass is given by [200]

$$R = \left(\sum_{i=1}^{N} \frac{m_i r_i^2}{M} \right)^{1/2} \qquad \text{with} \qquad M = \sum_{i=1}^{N} m_i . \qquad (12.1)$$

The multidimensional Schrödinger equation for the wavefunction $\Psi(R, \hat{\mathbf{R}})$ of N charged point-like particles takes a hydrogen-like form insofar as it separates into a "Coulomb" part and a kinetic energy, which in turn separates into an R-dependent and an $\hat{\mathbf{R}}$-dependent part. The theory for the multidimensional structure has been worked out by Fano et al. [200].

We here confine ourselves to the simplest three-body system, provided by the helium atom. Other three-particle aggregates are the H^- or the H_2^+ ion. In the former cases, the electron-electron interaction or the "electron correlation" plays a crucial role, in the latter, the ionic interaction. More generally, one may consider three-body systems with different masses of all the constituents. For details see the excellent review by C.D. Lin [201].

In order to illustrate the basic idea, I want to concentrate on the simple case of two electrons in the field of a nucleus with charge Z and infinite mass. If the electrons have the coordinates $\mathbf{r}_1 = (r_1, \hat{\mathbf{r}}_1) = (r_1, \theta_1, \varphi_1)$ and $\mathbf{r}_2 = (r_2, \hat{\mathbf{r}}_2) = (r_2, \theta_2, \varphi_2)$ with respect to a space-fixed coordinate system centered at the nucleus, we introduce the hyperspherical coordinates R and α by

$$R = \sqrt{r_1^2 + r_2^2} \qquad \text{and} \qquad \alpha = \arctan(r_2/r_1). \qquad (12.2)$$

Here, R is the "hyperradius" replacing R of Eq. (12.1) and α is the "hyperangle" contained in $\hat{\mathbf{R}}$. As a result, one has the replacement

$$(\mathbf{r}_1, \mathbf{r}_2) = (r_1, \theta_1, \varphi_1, r_2, \theta_2, \varphi_2) \longrightarrow (R, \Omega) = (R, \alpha, \theta_1, \varphi_1, \theta_2, \varphi_2). \tag{12.3}$$

Within the original set of coordinates, the two-electron Schrödinger equation in atomic units is given by

$$\left(-\frac{1}{2}\nabla_1^2 - \frac{1}{2}\nabla_2^2 - \frac{Z}{r_1} - \frac{Z}{r_2} + \frac{1}{r_{12}} - E \right) \tilde{\Psi}(\mathbf{r}_1, \mathbf{r}_2) = 0. \tag{12.4}$$

The distance between the electrons is denoted by r_{12} and the angle between their vectors with θ_{12}. The volume element in hyperspherical coordinates is found to be $d\mathbf{r}_1 d\mathbf{r}_2 = R^5 dR (\sin\alpha\cos\alpha)^2 d\alpha\, d\hat{\mathbf{r}}_1 d\hat{\mathbf{r}}_2$. It is convenient to eliminate the first-order derivatives in the differential operators of Eq. (12.4) by redefining the wavefunction as

$$\tilde{\Psi}(\mathbf{r}_1, \mathbf{r}_2) = \Psi(R, \Omega)/(R^{5/2}\sin\alpha\cos\alpha), \tag{12.5}$$

so that $\Psi(R, \Omega)$ satisfies

$$\left(-\frac{d^2}{dR^2} + \frac{\Lambda^2}{R^2} + \frac{2C}{R} + 2E \right) \Psi(R, \Omega) = 0, \tag{12.6}$$

which is of a hydrogen-like form. Here, in analogy to the usual separation in spherical coordinates, the first term is the kinetic-energy operator associated with the hyperradial motion and Λ^2/R^2 is the grand "centrifugal" potential energy operator including the single-particle centrifugal energy of each electron, where

$$\Lambda^2 = \left(-\frac{d^2}{d\alpha^2} + \frac{\mathbf{l}_1^2}{\cos^2\alpha} + \frac{\mathbf{l}_2^2}{\sin^2\alpha} \right) - \frac{1}{4} \tag{12.7}$$

is the square of the grand angular momentum operator with $\mathbf{l}_{1,2}$ being the angular momentum operators for the two electrons and C/R is the total Coulomb interaction potential among the three charged particles composed of the electron-nucleus and the electron-electron interactions, that is

$$C(\alpha, \theta_{12}) = -\frac{Z}{\cos\alpha} - \frac{Z}{\sin\alpha} + \frac{1}{\sqrt{1 - \sin 2\alpha \cos\theta_{12}}}. \tag{12.8}$$

The eigenfunctions $u_{l_1 l_2 m}(\Omega)$ of Λ^2 satisfy

$$\left(-\frac{d^2}{d\alpha^2} + \frac{\mathbf{l}_1^2}{\cos^2\alpha} + \frac{\mathbf{l}_2^2}{\sin^2\alpha} - (\nu + 2)^2 \right) u_{l_1 l_2 m}(\Omega) = 0, \tag{12.9}$$

where $\nu = l_1 + l_2 + 2m$ and $m \geq 0$ is an integer. For a given system, the total angular momentum L and its projection M have fixed values. The eigenfunctions can be decomposed as

$$u_{l_1 l_2 m}(\Omega) = f_{l_1 l_2 m}(\alpha)\, \mathcal{Y}_{l_1 l_2 LM}(\hat{\mathbf{r}}_1, \hat{\mathbf{r}}_2), \qquad (12.10)$$

with the coupled angular momentum function

$$\mathcal{Y}_{l_1 l_2 LM}(\hat{\mathbf{r}}_1, \hat{\mathbf{r}}_2) = \sum_{m_1 m_2} \left(\begin{array}{cc|c} l_1 & l_2 & L \\ m_1 & m_2 & M \end{array} \right) Y_{l_1 m_1}(\hat{\mathbf{r}}_1) Y_{l_2 m_2}(\hat{\mathbf{r}}_2).$$

$$(12.11)$$

Here, the $Y_{lm}(\hat{\mathbf{r}})$ are the usual spherical harmonics and $(\ |\ )$ is a Clebsch-Gordan coefficient. The eigenfunctions in α have the explicit form

$$f_{l_1 l_2 m}(\alpha) = N_{l_1 l_2 m} (\cos \alpha)^{l_1 + 1} (\sin \alpha)^{l_2 + 1}$$

$$\times\ _2F_1(-m, l_1 + l_2 + m + 2, l_2 + 3/2, \sin^2 \alpha), \ (12.12)$$

where $N_{l_1 l_2 m}$ is a normalization constant and $_2F_1$ a hypergeometric function. Since $-m$ is a negative integer, the series terminates and reduces to a Jacobi polynomial [202].

The total eigenfunctions $u_{l_1 l_2 m}(\Omega)$ are denoted as hyperspherical harmonics. They are harmonic functions on the hyperspherical surface of a six-dimensional space and are simultaneous eigenfunctions of the operators Λ^2, $\mathbf{l}_1^2$, $\mathbf{l}_2^2$, $\mathbf{L}^2$ and L_z. The set of hyperspherical harmonics satisfies orthogonality conditions

$$\int u_{l_1 l_2 m}(\Omega)\, u_{l_1' l_2' m'}(\Omega)\, \mathrm{d}\Omega = \delta_{l_1 l_1'} \delta_{l_2 l_2'} \delta_{mm'} \qquad (12.13)$$

and forms a complete set on the hyperspherical surface.

The nonseparability of the Schrödinger equation (12.6) is caused by the electron-electron interaction contained in the potential-energy term $2C/R$, where the "effective charge" $2C$, see Eq. (12.8), depends only on the angles α and θ_{12}. Figure 12.1 shows the potential surface on the (α, θ_{12})-plane for $Z = 1$.

The potential surface has the following important features: (a) The deep valleys near $\alpha = 0°$ and $\alpha = 90°$ reflect the strong Coulomb attraction if either electron is close to the nucleus. (b) The steep spike for $\alpha = 45°$ and $\cos \theta_{12} = 1$ corresponds to the strong Coulomb repulsion if the two electrons are very close to each other. (c) There is a saddle point at $\alpha = 45°$ and $\cos \theta_{12} = -1$. The potential drops if α either increases or decreases, but it

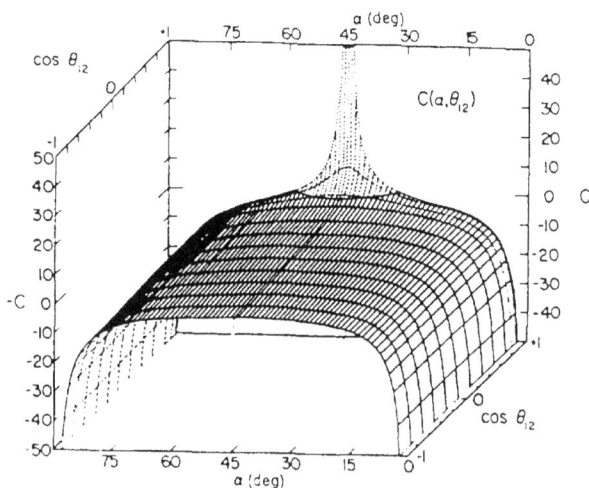

Figure 12.1: Potential surface $C(\alpha, \theta_{12})$ in Rydberg units (half of a.u.) for a pair of electrons in the field of an H^+ ion. From [201].

increases if θ_{12} increases. This potential point plays a major role for the quasi-stability of doubly excited states of atoms.

For a general three-body Coulomb system composed of particles A, B, C with different charges and masses, the hyperspherical approach has to be generalized correspondingly [201]. In a first step, Jacobi coordinates are introduced, e.g., the distance ρ_1 between A and B and the distance ρ_2 of C from the center of mass of A and B. If now μ_1 is the reduced mass of (A,B) and μ_2 of [(A,B),C] and μ an arbitrary reference mass, one has to modify Eq. (12.2) by the replacements $r_1 \rightarrow \sqrt{\mu_1/\mu}\,\rho_1$ and $r_2 \rightarrow \sqrt{\mu_2/\mu}\,\rho_2$. With these mass-weighted hyperspherical coordinates, the procedure follows the one outlined above.

12.2 Adiabatic approximation

The surge of applying hyperspherical coordinates owes much of its success to adopting the adiabatic approximation [203] in analogy to the treatment of slow collisions within the molecular orbital model, see Sec. 2.3. The wavefunction is written in the form of a Born-Oppenheimer expansion

$$\Psi(R,\Omega) = \sum_{\mu} F_{\mu}(R)\, \Phi_{\mu}(R,\Omega), \tag{12.14}$$

where the adiabatic wavefunctions $\Phi_{\mu}(R,\Omega)$ are obtained by solving the Schrödinger equation for a fixed value of the hyperradius R

$$H\,\big|_{R=\text{const}}\ \Phi_{\mu}(R,\Omega) = U_{\mu}(R)\Phi_{\mu}(R,\Omega)\,. \tag{12.15}$$

Similarly as in Secs. 2.4 and 2.8, it is assumed that $\Phi_{\mu}(R,\Omega)$ varies smoothly with R except in localized regions of avoided crossings.

Using the expansion (12.14), the adiabatic function $\Phi_{\mu}(R,\Omega)$ satisfies the equation

$$\left(\Lambda^2/R^2 + C/R\right) \Phi_{\mu}(R,\Omega) = U_{\mu}(R)\, \Phi_{\mu}(R,\Omega), \tag{12.16}$$

where Λ^2 is given by (12.7) and C by (12.8) while the hyperradial functions satisfy the coupled equations

$$\left(-\frac{d^2}{dR^2} + U_{\mu}(R) - 2E\right) F_{\mu}(R)$$

$$= \sum_{\nu} \left(2\langle \Phi_{\mu} \mid \frac{d}{dR} \mid \Phi_{\nu}\rangle \frac{d}{dR} + \langle \Phi_{\mu} \mid \frac{d^2}{dR^2} \mid \Phi_{\nu}\rangle\right) F_{\nu}(R). \tag{12.17}$$

The adiabatic approximation is useful, if the off-diagonal coupling terms are small. In this case, the index μ can be used to label the "channels". While the hyperradial function gives the size of the state, the internal motion, together with the overall motion of the whole system, is contained in the channel function. One of the main goals in the study of three-body or few-body systems is to identify the different modes of internal motion. They provide meaningful quantum numbers represented by the channel index μ.

12.3   The hyperspherical close-coupling method and applications

The numerical solution of the coupled equations (12.17) suffers from the existence of nonadiabatic coupling terms, which vary rapidly near avoided crossings, so that the resulting accuracy with a finite basis set is quite limited. A serious limitation of the adiabatic expansion (12.14) for reactions arises from the incorrect asymptotic form. At large hyperradii, there are several ways into which a three-body system can dissociate. For any given disintegration channel, there are nonvanishing coupling terms in the asymptotic region. For example, the nonadiabatic coupling term $\langle \Phi_\mu | \mathrm{d}/\mathrm{d}R | \Phi_\nu \rangle$ in Eq. (12.17) decreases as $1/R$ for large R.

These difficulties are circumvented in the hyperspherical close-coupling method. In this approach, the range of the hyperradius is first partitioned into small segments $\sigma = 1, 2, \cdots$ around values R_σ. Within each segment, the wavefunction is expanded as

$$\Psi(R, \Omega) = \sum_\mu F_\mu(R)\, \Phi_\mu(R_\sigma, \Omega). \tag{12.18}$$

The coupling terms are smooth functions of R within a segment. In this way, each segment is treated separately, and the logarithmic derivatives of the wave functions $F_\mu(R)$ are propagated from one segment to the next. In the asymptotic region, the solutions propagated so far, are matched to the desired asymptotic solutions expressed in the appropriate Jacobi coordinates.

The hyperspherical close-coupling method has been widely applied, usually to low collision energies. A generalization of the method has been proposed and demonstrated for the charge transfer in He^{2+} + H(1s) collisions at low energies [204]. Another example with two leptons is the unified treatment of positron annihilation and positron formation in scattering with a hydrogen atom [205]. Reactions for collision systems with arbitrary masses, for which the proper Jacobi coordinates have to be taken into account, include elastic and spin-flip processes in $p + p\mu$ and $e^{\pm}+$Ps [Ps=(e^+e^-)] scattering [206], muon transfer between hydrogen ions [207] and low-energy rearrangement processes in the $dt\mu$ system [208] (where d denotes the deuteron and t the tritium) in connection with the problem of muon-catalyzed fusion, to H(1s) + μ^+ rearrangement collisions. More recently, two-body correlations in Bose-Einstein condensates [209] as well as antiproton-hydrogen collisions [210] have been studied.

These are just a few examples in the vast literature on the application of hyperspherical close-coupling methods. In all cases considered, the collision energy is very low. For collisons between highly-charged heavy ions, it is usually not necessary to treat all three (or more) particles on the same footing and hence resort to the hyperspherical methods.

Chapter 13

Selected Topics II: Hollow atoms in micro-capillaries

In this Chapter, we briefly discuss collisions of low-energy multicharged ions with metal and other surfaces, a subject that has found great interest because of the exotic nature of collision dynamics and the possible application to surface analysis and surface modification. For a recent review, see [4]. After a qualitative discussion of the dynamics in ion-surface collisions leading to "hollow atoms" in Sec. 13.1, we study in Sec. 13.2 the effect of mirror images of point charges in front of a metal surface and the resulting potential energy surfaces. In Sec. 13.3 we consider the passage of ions through micro-capillaries in metal foils and in Sec. 13.4 in polymer foils, which offer the possibility of guiding the ion beam away from its original direction.

13.1   Formation of hollow atoms

When a highly-charged ion approaches a metal surface, it is accelerated toward the surface by the interaction with its image charge and, shortly before plunging into the surface, it captures target electrons into excited states, mainly into outer shells with radii much larger than those of the shells initially occupied. As a result, a "hollow atom" is formed, in which outer shells are populated while inner shells still remain vacant, see Fig. 13.1. Subsequently, the system shrinks by Auger processes, in which outer-shell electrons are de-excited, while transferring the gained energy to ionized electrons. Nevertheless, the time interval between the formation of a hollow atom above the surface (denoted as HA1) and its arrival at the surface is less than $10^{-14} - 10^{-13}$ sec, that is, shorter than its natural lifetime [211].

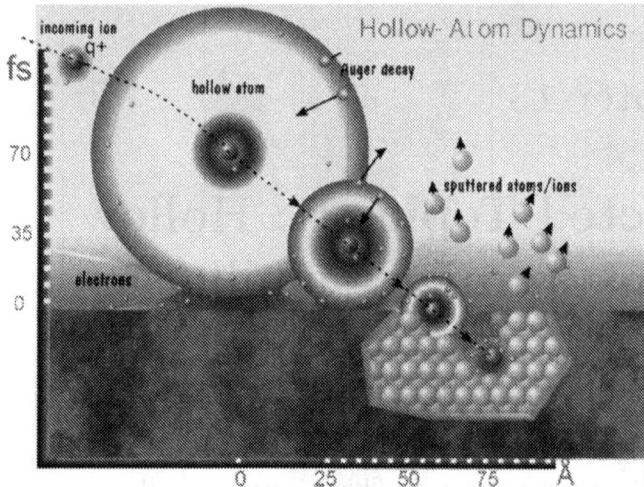

Figure 13.1: Visualization of the formation of hollow atoms. The horizontal and the vertical axes give scales in distance and in time, respectively. From [212].

As the ion penetrates the surface, target electrons dynamically screen the ion charge, promote ionic energy levels, so that some upper inner-shell vacancies can be filled via quasiresonant charge transfer and Auger transitions. However, if an incident ion has vacancies in deeper inner shells, a hollow atom can still persist at or below the surface. This hollow atom of the second generation (denoted as HA2) eventually decays by emitting x-rays and/or Auger electrons. In order to suppress HA2 and single out HA1, thin foils with straight micro-capillaries have been employed as targets, see, e.g., [213], so that some HAI can pass through the foil without further interaction. Highly-ordered micro-capillary foils of Ni and Au have been fabricated with sophisticated nano-lithograpic techniques [214].

In the following, we discuss only the simplest, yet most important theoretical considerations without going into more complicated details of solid-state physics.

13.2   The classical over-barrier model

In order to understand the basic mechanism of ion-surface processes, we represent the surface by an infinite unstructured plane in the x, y direction and a one-dimensional potential in the z direction perpendicular to the plane. This is usually called the "jellium model", in which, for a metal, the conduction band reaches up to the Fermi energy E_F, separated by the energy W, the "work function", from the potential $V = 0$ at infinity. The potential of the metal surface with an ion in front of it, is illustrated

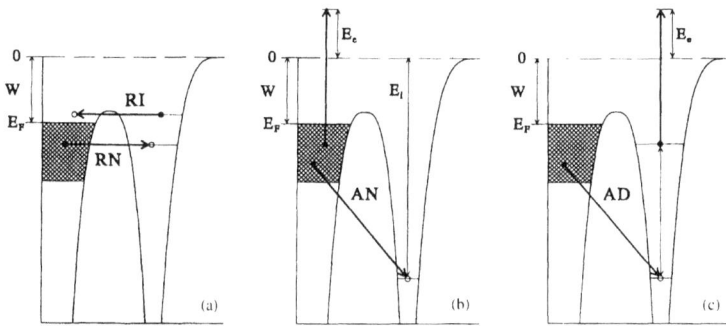

Figure 13.2: Schematic illustration of various types of charge transfer processes at a metal surface: (a) Resonance neutralization (RN) and resonance ionization (RI); (b) Auger neutralization (AN); (c) Auger de-excitation (AD). From [4].

schematically in Fig. 13.2. We are mainly interested in panel (a), in which resonance neutralization requires the metal electrons to tunnel through the potential barrier between the surface and the ion. However, as the ion approaches the surface, the barrier is lowered, so that from a certain distance on, the electrons can enter the ions *above* the barrier. This process is classically allowed and hence has a much higher probability to occur than tunneling. The "classical over-barrier model" has been employed in various connections within atomic collision physics and has been adopted to ion-surface processes by Burgdörfer et al. [215, 216, 217].

For a quantitave treatment of the over-barrier model, we condider the geometry in more detail. Let $\mathbf{R}$ and $\mathbf{r}_e$ be the positions of the nucleus and the electron with respect to the projection point of the ion on the surface, respectively. Choosing the y direction perpendicular to the plane spanned by $\mathbf{R}$ and $\mathbf{r}_e$, we have $\mathbf{R} = (0, 0, d)$ and $\mathbf{r}_e = (x, 0, z)$, where d and z are the distances of the ion and the electron from the surface, respectively. Both charges produce image charges behind the surface. This is illustrated in Fig. 13.3 (a).

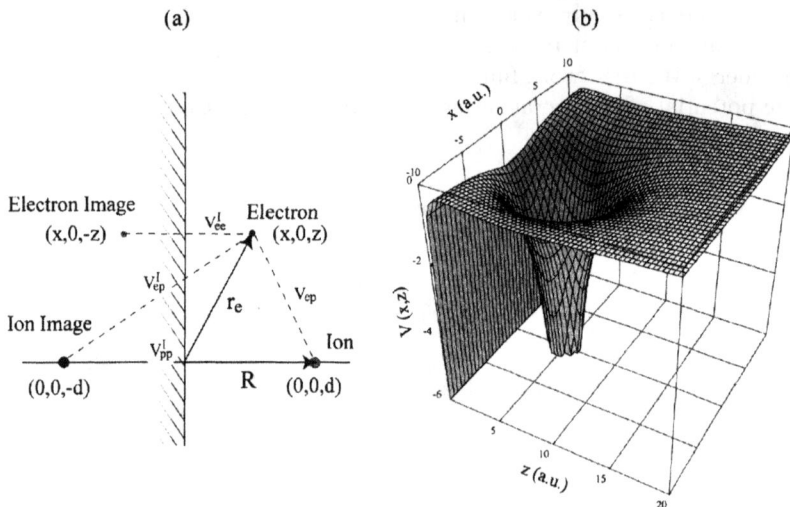

Figure 13.3: Interactions of the active electron with the ion and the metal surface: (a) Ion and electron in front of the metal surface (b) Three-dimensional representation of the potential $V(\mathbf{r_e}, \mathbf{R})$ acting on the electron for a nuclear charge $q = 10$ and the distance $d = 5$ a.u. from the surface. From [218].

We first consider the interaction of a point charge q at distance z from the surface with its own image charge $-q$ at the position $-z$. The potential is derived by integrating the attractive force between the charge and its image from infinity to the position z. While for a fixed position of one of the charges, the potential is proportional to the inverse distance, the change of the image charge's position during the approach leads to a reduction by a factor $1/2$. From the force $-q^2/(2z)^2$ between q and its image, we obtain the classical potential energy

$$V_{\text{im}}(z) = -\int_z^\infty \frac{q^2}{(2z')^2} \mathrm{d}z' = -\frac{q^2}{4z}. \tag{13.1}$$

The interaction of the electron with the ion, with the ion's image and with its own image is given (in atomic units) by

$$V(\mathbf{r_e}, \mathbf{R}) = -\frac{q}{|\mathbf{r_e} - \mathbf{R}|} + \frac{q}{|\mathbf{r_e} + \mathbf{R}|} - \frac{1}{4z}. \tag{13.2}$$

This potential is depicted in Fig. 13.3 (b). It is noted that a saddle point appears on the line connecting the ion and its image. In order to determine its position and height, we take $x = 0$ and $0 < z < d$ in Eq. (13.2), so that

$$V(z, d) = -\frac{q}{d - z} + \frac{q}{d + z} - \frac{1}{4z}. \tag{13.3}$$

By forming the derivative with respect to z, setting it equal to zero, and by assuming $d^2 \gg z^2$, we obtain the position of the saddle point as

$$z_0 = \frac{d}{\sqrt{8q}}. \tag{13.4}$$

Since we are interested in multiply charged ions, we have indeed $z_0^2/d^2 = 1/8q \ll 1$ as assumed before. The barrier height is obtained by inserting z_0 into Eq. (13.3) and expanding up to linear terms in $1/\sqrt{8q}$, so that

$$V_{\text{saddle}} = -\frac{\sqrt{2q}}{d}. \tag{13.5}$$

Now equating $|V_{\text{saddle}}|$ with the work function W, see Fig. 13.1 (a), we derive the critical distance of the ion from the surface for over-barrier capture to be possible, that is,

$$d_c = \frac{\sqrt{2q}}{W}, \tag{13.6}$$

as employed in [211, 218].

The principal quantum number n_c into which capture occurs at the top of the barrier is given by the ionic quantum number at the saddle point, supplemented by an energy shift induced by the image charge [215]:

$$n_c^2 = \frac{q^2}{2(W + q/2d_c)} = \frac{q^2}{2W\left(1 + \sqrt{q/8}\right)} \tag{13.7}$$

and

$$n_c = \frac{q}{\sqrt{2W}} \frac{1}{\sqrt{1 + \sqrt{q/8}}}. \tag{13.8}$$

Clearly, the actual situation is much more complicated and there are sophisticated theortical treatments available, see, e.g., [4, 215, 216, 217]. However, even with these treatments, some questions are still open, and it remains a further challenge to handle the ion-surface processes theoretically. Meanwhile, for an analysis of experiments, the basic relations presented here, are indispensible and in many cases sufficient.

13.3   Passage through micro-capillaries in metals

As has been outlined above, in usual glancing collisions of slow highly-charged ions, the lifetime of hollow atoms of the first generation (HA1) is very short $(10^{-14} - 10^{-13}$ sec) before they plunge into the surface, continue as modified hollow atoms of the second generation (HA2) and soon afterwards collapse. The nature of the original hollow atoms (HA1) can only be studied when extracted and studied in vacuum.

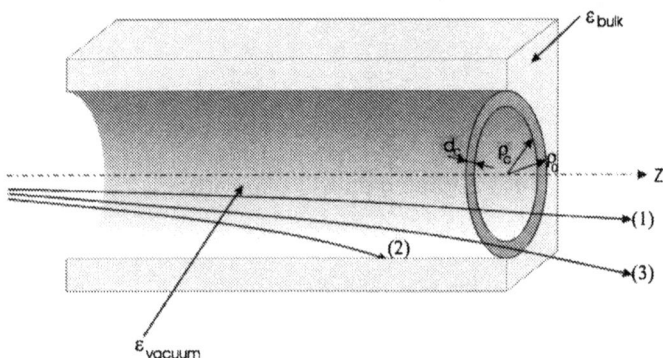

Figure 13.4: Illustration of an ideal micro-capillary with typical ion trajectories. Here, ρ_0 is the inner radius of the capillary, ρ_c the critical capture radius and d_c the capture distance (13.6). From [217].

In order to investigate the formation process and the nature of HA1, foils of straight micro-capillaries have been used, into which multicharged ions have been shot [211, 214, 219, 220, 221, 213]. Depending on the initial position of the ion trajectory with respect to the wall of the capillary, the ions may suffer three different fates, see Fig. 13.4 and [217, 213].

(1) If the distance of the ion to the inner wall is always larger than the critical distance $d_c(q)$, the ion passes through the capillary while keeping its initial charge state. (2) If the distance at some time becomes shorter than d_c, a hollow atom HA1 is formed by resonant charge transfer, subsequently it hits the inner wall and collapses. (3) If, finally, HA1 is formed near the exit of the capillary, it can escape into the vacuum with a charge different from the initial one. The ratio between the events (3) and (1) can be estimated classically as the ratio $f(q)$ of the area spanned by the ring

between ρ_c and ρ_0 compared to the cross section of the capillary, that is

$$f(q) = 2\pi\rho_c d_c(q)/\pi\rho_c^2 = \sqrt{8q}/\rho_c W. \tag{13.9}$$

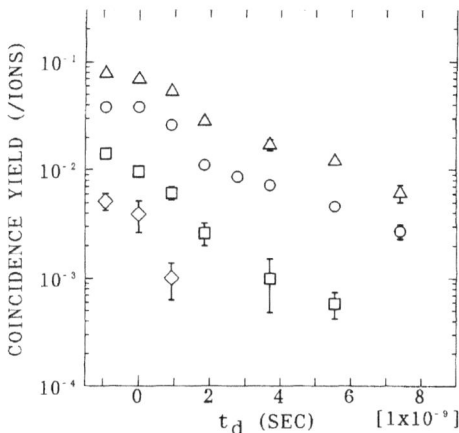

Figure 13.5: Integrated delayed x-ray yields produced by 2.1 keV/u Ni^{6+} ions transmitted through Ni micro-capillaries in coincidence with the detected charge state q (normalized per one N^{q+} ion). Triangles: $q = 5$; circles: $q = 4$; squares: $q = 3$; diamonds: $q = 2$. From [211].

With a critical distance d_c of the order of nm and a typical capillary diameter of 100 nm, one obtains a fraction for charge transfer of about 1%, which is in accord with experimental findings. Of course, there will be multiple charge transfers. However, for the second charge transfer, it is no longer justified to use the approximation of a point charge q as in Sec. 13.2, because the resonantly captured electron sits in an outer shell. Furthermore, autoionization will influence the final charge distribution measured behind the foil. A theoretical description is, therefore, very difficult.

Besides measuring the charge states of the ions emerging behind the Ni capillary foil, one may investigate the early stages of HA1 formation by visible light spectroscopy. For example, with ion beams of 2 keV/u Ar^{q+} ($q = 6 - 11$) transmitted through a Ni capillary, strong lines in the visible spectrum typically for principal quantum numbers $n \approx q + 1$ occur, in accord with the over-barrier model. By high-resolution measurements, one can get information on the initial-state distribution.

An interesting feature is revealed in the analysis of the late stages of
HA1 formation and the shell-filling process by detecting the soft x-rays
emitted from ions after transmission through a capillary. While in glancing
scattering from a surface, multicharged ions are found to be neutralized
within about 30 fs, the transmission through capillaries leads to lifetimes
that are longer by several orders of magnitude. As an indicator, one may
use the delayed integrated x-ray yield, see Fig. 13.5. The x-rays have been
measured [211] in coincidence with the charge state of the emerging ion.
Clearly, in the time span between the x-ray emission and the charge-state
detection, the ions have relaxed and the charge state has been increased by
the emission of Auger electrons, that is, the charge state q given in Fig. 13.5
is equal or larger than at the time of photon emission. However, it is seen
that the decay curve is rather independent of the charge state q.

The measured decay times of hollow atoms turn out to be in the range
of a few ns. Hence, the passage through capillaries stabilizes hollow atoms
to lifetimes that are about six orders of magnitude longer than in glancing
collisions with a surface.

13.4   Capillary guiding in polymer foils

In Section 13.3 we have discussed the passage through highly-ordered capillaries produced in Ni foils by nano-lithographic techniques [214]. Another
way to create capillaries with great precision, are ion tracks caused by energetic projectiles [222], for example in polymer foils. Using well-known
etching techniques, the ion tracks can be converted into capillaries with
mesoscopic diameters ranging from a few nm to a few μm [223].

Stolterfoht et al. [224] have studied transmission of Ne^{7+} ions through
capillaries of 100 nm diameter formed by etching ion tracks in highly insu-
lating foils of polyethylene terephthalate (PET or mylar) with a thickness
of 10 μm. As a novel feature, the angular distributions provide evidence
that the Ne^{7+} ions are guided through the capillaries. This is a remarkable
effect for highly charged ions, because it involves multiple scattering events
at the inner capillary surface while preserving the incident charge state.

The guidance of the ions becomes apparent, when the PET foil is tilted
by an angle of $\psi = 0°$ or $5°$ and the maximum ion intensity behind the
foil is observed at an angle $\theta = \psi$, that is exactly in the direction of the
capillary. Figure 13.6 shows ion spectra obtained for an observation angle
equal to the tilt angle. It is seen that the intensity of the transmitted ions
with the initial charge state 7 is dominant while the intensity of the other
charge states is by orders of magnitude smaller.

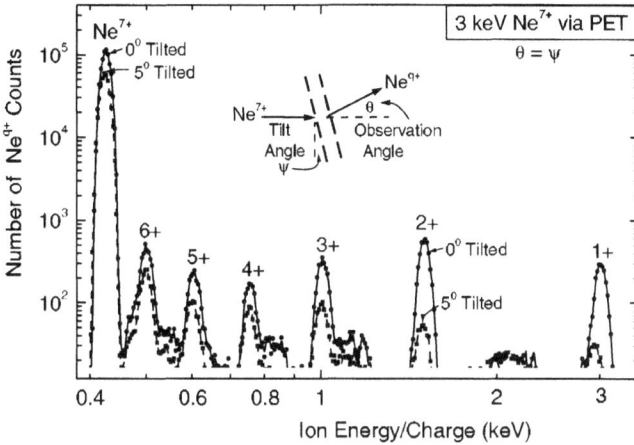

Figure 13.6: Charge-state spectra of Ne^{q+} ions produced after the passage of 3 keV Ne^{7+} through capillaries in PET. The charge state q is given at each peak. The tilt angles are $\psi = 0°$ and $5°$. The ion observation angle is $\theta = \psi$ [224].

In Fig. 13.7, the integrated peak intensity of the transmitted Ne^{7+} ions is plotted as a function of the observation angle θ for different tilt angles ψ. On the other hand, for capillaries coated with Ag, the angular distribution is very narrow and centered around $\theta = 0°$. Already for a tilt angle of $\psi = 5°$, there is almost no intensity. The most important finding is that the direction of the incident Ne^{7+} ions is altered to a preferential direction parallel to the capillary axis. Since the aspect ratio between the length and the diameter of the capillary is 100, the maximum deflection for a straight trajectory through the capillary is $\approx 0.5°$, consistent with the Ag data. Consequently, Ne^{7+} ions transmitted through foils with angles $\geq 5°$ have to interact at least once with the inner wall of the capillary. It should be noted that the Ne^{7+} intensity is still significant at tilt angles a large as $\psi = 20°$.

The enhanced ion transmission suggests that the inner walls of the capillaries collect charges so that the build-up of electrostatic repulsion inhibits further close collisions with the surface, thus avoiding electron capture by the projectile. This can be understood if the incident ions initially deposit positive charges on the capillary surface until the ions are deflected by electrostatic repulsion. As a result, according to Fig. 13.7, the ion beam is

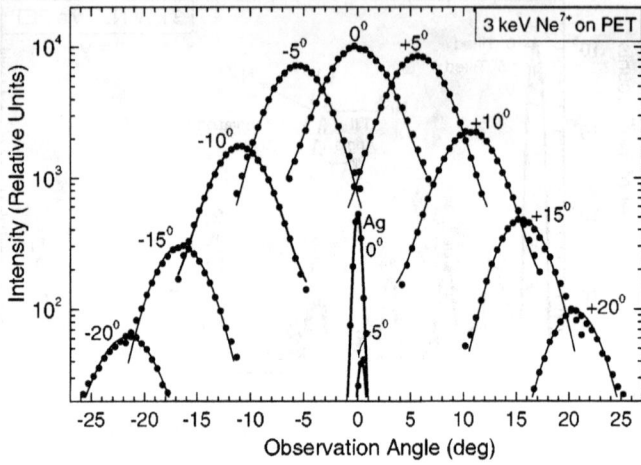

Figure 13.7: Angular distributions of Ne^{7+} ions transmitted through capillaries in PET. The foils were tilted by the angles indicated in the figure. The solid lines represent Gaussian functions fitted to the data. The narrow peaks near $0°$ were obtained using capillaries covered by Ag. From [224].

guided with an angular divergence of about $\pm 2.5°$.

Chapter 14

Selected topics III: Resonant coherent excitation

In Chapter 13, we have discussed the passage of low-energy multicharged ions through micro-capillaries in metals and polymers, which were treated as unstructured media. In the present Chapter, we take the lattice structure into account and consider collisions of high-energy ions with single crystals. Energetic ions travelling through a crystal along one of its axes or planes are trapped in the static potential well formed by atomic strings or planes. The ions then are guided by "channels" in the crystal and experience less scattering and energy loss than ions incident on the crystal in arbitrary directions. An ion moving within a channel inside a crystalline solid experiences a variation of the static electric field as it passes each lattice atom [225]. The variation of the field along the trajectory appears in transmission channelling as a periodic time-varying electric field in the projectile frame. The fundamental perturbation frequency ν experienced by the ion is $\nu = v/d$, where v is the ion velocity and d is the distance of atoms in a row (or string). Since the induced time-dependent field is not purely sinusoidal, also integral multiples of the fundamental frequency will occur in the Fourier expansion [226, 227].

In Sec. 14.1, we briefly introduce the axes and planes in a single crystal, then proceed in Sec. 14.2 to the excitation process affecting an ion during the passage and in Sec. 14.3 to its detection. Finally, in Sec. 14.4, the influence of the ion's initial impact parameter with respect to the channel axis on its energy loss is studied.

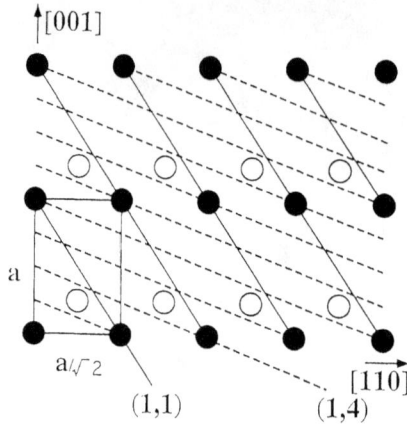

Figure 14.1: Atomic arrangement in a $(2\bar{2}0)$ plane of the diamond structure. The two-dimensional basis vectors are $\mathbf{A} = [110]a/\sqrt{2}$ and $\mathbf{B} = [001]a$. The solid and dashed lines represent (1,1) and (1,4) strings, respectively. From [228].

14.1 Axes and planes in a crystal

In order to describe the excitation process, we first have to specify the directions in a crystal in which a periodic excitation can occur. These special directions and, similarly, planes are characterized by the "Miller indices". For example, in a cubic crystal, the basic x, y, z *directions* are indicated by [100], [010], and [001], respectively. The diagonal in the $x - y$ plane is given by [110], and the space diagonal in a cube by [111]. The *plane (klm)* is orthogonal to the *direction [klm]* in a cubic crystal. For a more complicated lattice, the specification of an axis or of a plane is quite involved and requires a 3-dimensional visualization.

One such example is the diamond structure of Si crystals, in which we are mainly interested. We consider the $(2\bar{2}0)$ plane which is perpendicular to the $x - y$ plane and intersects the x axis at $x = 2$ and the y axis at $y = -2$. The axis [110] is the intersection of the $x - y$ plane with the $(2\bar{2}0)$ plane. If a is the lattice constant of an fcc crystal, the basis vectors in the $(2\bar{2}0)$ plane are $[110]a\frac{1}{2}\sqrt{2}$ and $[001]a$, see Fig. 14.1. Within this plane, the strings of atoms are characterized by (k, l), where k and l are integers. In Fig. 14.1 the strings (1,1) and (1,4) are also indicated.

14.2   The excitation process during the passage

When one of the frequencies felt by the ion travelling along a crystal axis or in a crystal plane corresponds to the difference of internal energies of the ion, *resonant coherent excitation* (RCE) will occur. The RCE process and the possibility of radiative de-excitation has been predicted originally by Okorokov in 1965 [225] and has been first observed by Datz et al. in 1978 [226] via the charge state distribution of the transmitted ions. Since then, much progress has been made in experiment and theory, see [227]. Most of the early experiments have been performed with hydrogen-like light ions such as F^{8+} and Ne^{9+} leading to the excitation of $n = 2$ states in the projectiles with energies of several MeV/u.

More recently, experiments with relativistic 390 MeV/u hydrogen-like Ar^{17+} and Fe^{25+} as well as with He-like Ar^{16+} and Fe^{24+} have been performed at the Heavy Ion Medical Accelerator in Chiba, Japan, (HIMAC) by Azuma et al. [228, 229, 230, 231, 232]. In the following, we discuss these experiments and defer an analysis of the advantages to Sec. 14.3. We here just state the resonance condition, which is modified for relativistic collisions by replacing the lattice constant a with the Lorentz-contracted form a/γ, where $\gamma = (1 - \beta^2)^{-1/2}$ and $\beta = v/c$. Let θ be the angle between the beam direction and the [110] axis, k and l integers, then the resonance is achieved for

$$\frac{k \cos \theta}{A} + \frac{l \sin \theta}{B} = \frac{\Delta E}{\gamma \beta h c}, \tag{14.1}$$

where A and B are the effective periodicity lengths in the [110] and [001] directions, respectively. In the case of the $(2\bar{2}0)$ plane illustrated in Fig. 14.1, the lengths are $A = a/\sqrt{2}$ and $B = a$. For other channelling planes, A and B assume different values. For a given collision energy and a given pair of integers (k, l), Eq. (14.1) implies a one-to-one correspondence between a transition energy ΔE and a tilt angle θ with respect to the [110] direction. The perturbing lattice potentials acting on the relativistically moving projectile system can be obtained from Eq. (5.32) in Sec. 5.5

14.3   Detection of the excitation

In order to meet the channelling condition and to employ Eq. (14.1) for spectroscopy of hydrogen-like excited states, one needs a well-collimated beam (angular divergence < 0.15 mrad) and a precision 3-axes goniometer. Detection of RCE is achieved essentially by two methods

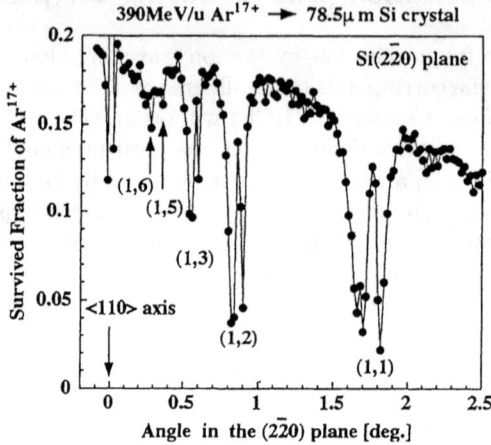

Figure 14.2: The surviving fraction of 390 MeV/u Ar^{17+} ions at various crystal axes (k,l) as a function of the tilt angle between the ion velocity and the [110] axis in the $(2\bar{2}0)$ plane. From [228].

- Because the $n = 2$ states in Ar^{17+} are spatially more extended than the 1s ground state, they are more easily ionized during their passage through the lattice. Hence, at the resonance, one observes a dip in the fraction of surviving Ar^{17+} ions, see Figs. 14.2 and 14.3.

- One may detect the de-excitation x-rays corresponding to $n = 2 \rightarrow n = 1$ transitions.

There are several advantages of relativistic collisions compared to the MeV/u energy range. (a) The dominant first-order component of the periodic crystal potential leads to large excitation probabilities. (b) The contribution of electron capture is very small, see Chapter 4, in particular Sec. 4.8, so that the RCE process is not obscured by capture. (c) The modification of the resonances by the Stark effect is caused *only* by the transverse crystal field, *not* by the dynamically induced wake field. The latter plays an important role in the lower energy regime but is negligible for relativistic channelling.

Figure 14.2 shows an overview over a wider angular range exhibiting characteristic dips in the survival fraction of Ar^{17+} ions at various crystal axes (k,l) entering in Eq. (14.1). Specifically, Fig. 14.3 displays the case

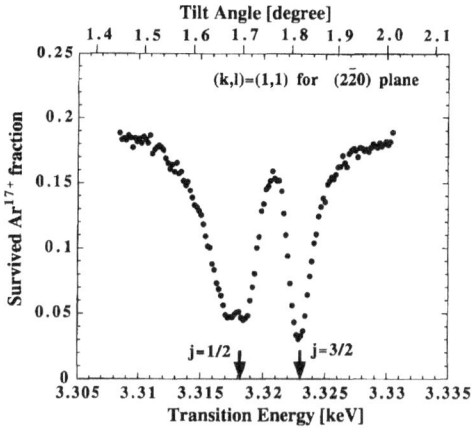

Figure 14.3: The RCE peak of $(k,l) = (1,1)$ in $(2\bar{2}0)$ planar channelling. The surviving Ar^{17+} fraction passing through a 94.7-μm Si crystal is plotted as a function of the tilt angle from the [110] axis (upper scale) and as a function of the transition energy (lower scale). The two arrows indicate the resonance positions of the 1s-2p energy levels of Ar in vacuum including the Lamb shift. From [230, 231].

of $(k,l) = (1,1)$ in more detail. The dips can be assigned to $j = 1/2$ and $j = 3/2$ states of Ar^{17+} (including the Lamb shift). All four of the $n = 2$ substates are subject to Stark mixing caused by the crystal potential. In particular, the $j = 1/2$ level is subject to a broadening and an indication of splitting, which is associated with the admixture of the 2s state.

In Figure 14.4, we present a comparison of the two detection methods applied to the same transition. The lower figure (b) gives the Ar^{18+} fraction corresponding to excitation, while the upper figure (a) exhibits the de-excitation x-rays measured in different directions and indicating an anisotropy.

14.4 Impact-parameter dependence

An important parameter which is indirectly accessible to measurement is the impact parameter of the projectile trajectory with respect to the center plane in planar channelling [231]. Clearly, one cannot measure the impact parameter directly, but one can measure the energy loss, which is uniquely

Tilt angle [degree]

1.4   1.5   1.6   1.7   1.8   1.9   2.0

Si(Li)$_v$

Si ($2\bar{2}0$)

Ar^{17+}

Si(Li)$_h$

(a)

● horizontal
○ vertical

j=1/2   j=3/2

X-ray yield (x-rays/ion)

0.3   0.2   0.1   0

Ar^{18+} fraction

0.8   0.7   0.6

(b)

3.305   3.31   3.315   3.32   3.325   3.33

Transition energy [keV]

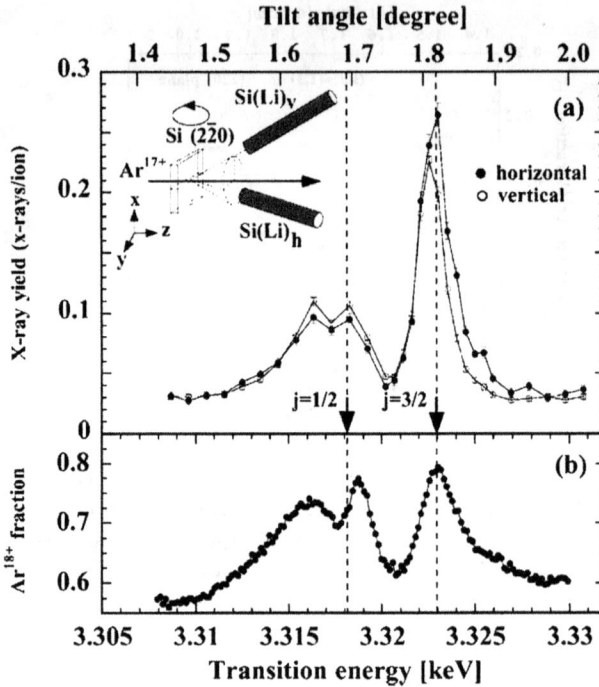

Figure 14.4: RCE profiles of $(k,l) = (1,1)$ transitions in $(2\bar{2}0)$ channelling through a 21 μm thick Si crystal as a function of the tilt angle between the beam direction and the [110] axis. (a) de-excitation x-rays at a horizontal (open circles) and a vertical (full cricles) position of the detector; (b) ionized Ar^{18+} fraction in transmitted Ar ions. From [232]

related to the ion trajectory, as can be established by simulations [229]. With increasing distance of the initial trajectory from the center of the channel, the amplitude of the oscillation increases and hence the energy loss.

Figure 14.5 shows a contour map of the ionized fraction Ar^{18+} through RCE as a function of the tilt angle or, equivalently, as a function of the excitation energy, and the dependence on the energy deposition (equivalent to the trajectory amplitude). The parameters are the same as in Fig. 14.3. At the lowest energy deposition, the RCE probability exhibits two narrow peaks for the $j = 1/2$ and $j = 3/2$ state of Ar^{17+}. As the amplitude

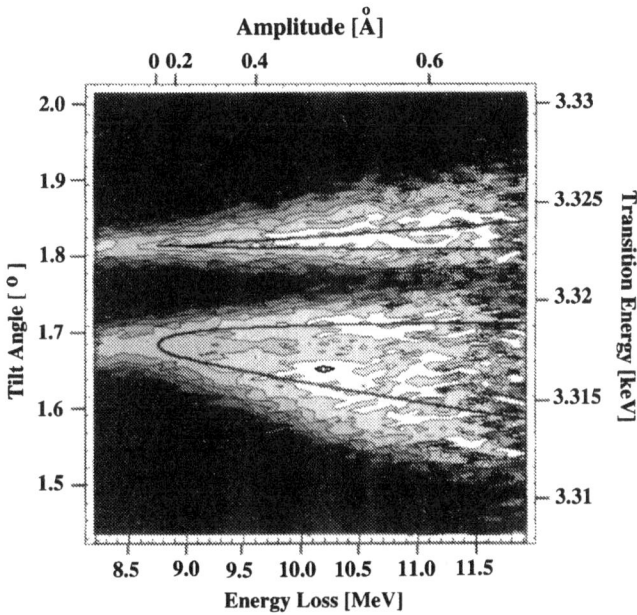

Figure 14.5: Contour map of the fraction ionized by RCE as a function of the tilt angle (equivalent to transition energy) and energy deposition (equivalent to trajectory amplitude), From [228].

of the trajectory increases (and hence the initial impact parameter), the higher transition energy broadens without shifting, while the lower transition energy splits into two peaks. These features are well reproduced by the calculated lines.

Besides the impact-parameter dependence, also the anisotropy in the emission of the de-excitation x-rays [232], see Fig. 14.4, and the emission of convoy electrons have been measured.

Chapter 15

Selected topics IV: Atomic physics with antiprotons

Collisions of low-energy antiprotons ($\bar{p}$) with matter occupy a unique place in atomic reaction studies. In order to explore this new field, an experimental facility, denoted as Antiproton Decelerator (AD), replacing the older LEAR, has been built in an international cooperative at CERN devoted to atomic collisions with slow antiprotons. At a later stage, studies with cold antiprotons will be resumed at the future Facility for Antiproton and Ion Research (FLAIR) at the GSI Darmstadt. Hence, in the near future, it will be possible to conduct precision experiments with *negative* structureless heavy ions. One of the goals is to produce significant numbers of cold antihydrogen atoms ($\bar{H}$) in order to test fundamental symmetries in physics. For example, an accurate spectroscopy of electromagnetic transitions in antihydrogen and a comparison with the corresponding transitions in hydrogen could answer the question whether the Rydberg constant in both systems are identical as required by the CPT theorem. One would also like to check whether matter and antimatter behave the same way under gravitational forces. These are all extremely difficult experiments. For a summary, see, e.g. [233].

Several strategies have been discussed to produce cold antihydrogen:

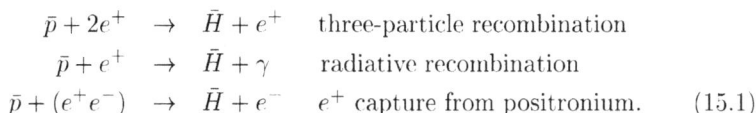

$$\bar{p} + 2e^+ \quad \rightarrow \quad \bar{H} + e^+ \quad \text{three-particle recombination}$$
$$\bar{p} + e^+ \quad \rightarrow \quad \bar{H} + \gamma \quad \text{radiative recombination}$$
$$\bar{p} + (e^+e^-) \quad \rightarrow \quad \bar{H} + e^- \quad e^+ \text{ capture from positronium.} \quad (15.1)$$

In all cases, a third body is needed on the right-hand side in order to satisfy energy-momentum conservation. Rather than dwelling on these production processes, we here wish to discuss the analog to ion-atom collisons, starting

223

in Sec. 15.1 with antiproton collisions with hydrogen atoms. In Sec. 15.2, we study the formation of antiprotonic helium, which has received much attention in the last years.

15.1   Collisions of antiprotons with hydrogen atoms

When a slow antiproton collides with a hydrogen atom, there are only two reaction channels open. Depending on the collision energy, the antiproton can replace the electron in the atom or it can simply ionize the hydrogen atom.

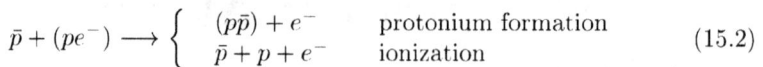

$$\bar{p} + (pe^-) \longrightarrow \begin{cases} (p\bar{p}) + e^- & \text{protonium formation} \\ \bar{p} + p + e^- & \text{ionization} \end{cases} \qquad (15.2)$$

If the antihydrogen energy in the center-of-mass system is less than the ionization threshold of 13.6 eV, protonium formation [Pn=$(p\bar{p})$] becomes dominant, where protonium, in analogy to positronium, is a bound state of a proton and an antiproton. More specifically, capture of $\bar{p}$ by hydrogen, $\bar{p} + H(1s) \rightarrow Pn(nl) + e^-$ occurs most likely via an energy matching or resonance condition [234] with

$$n \approx \sqrt{\mu_{p\bar{p}}/m_e} \approx 30, \qquad (15.3)$$

where n is the initial quantum number at the time of capture, before the antiproton cascades down into lower states, and $\mu_{p\bar{p}}$ denotes the reduced mass of $p\bar{p}$. For 10 eV antiprotons, the most probable initial population is estimated to be in the range of $n=$ 59 60 [210]. After capture into the n shell, the decay occurs via radiative or nonradiative pathways. It is usually thought that capture takes place into circular orbits $l = n - 1$ or almost circular orbits, because antiprotons in low angular-momentum states are subject to rapid annihilation.

The theoretical treatment has been performed in hyperspherical coordinates, see Chapter 12, [235, 210], which allows one to contract the set of coordinates into a single radial coordinate, the hyperradius (12.2) and into angular coordinates on the hypersphere. It is found theoretically that a protonium negative ion does not exist in the nonrelativistic limit and that no excited resonant states of the negative ion ($Pn\,e^-$) are expected for principal quantum numbers $n < 30$. Fully quantum mechanical calculations are also available [236]. So far, protonium formation has not been observed experimentally.

For collision energies exceeding 1 keV, $p\bar{p}$ formation is negligible [237], and one has to consider only the ionization channel. In this case, coupled-

channel calculations are applicable. Adopting a classical straight-line tra-
jectory $\mathbf{R} = \mathbf{b} + \mathbf{v}t$ for the antiproton motion and denoting with $\mathbf{r}$ the
position of the electron with respect to the proton, the time-dependent
Schrödinger equation is written as

$$\left[h(\mathbf{r}) + V(\mathbf{r}, \mathbf{R}(t)) - \mathrm{i} \left(\frac{\partial}{\partial t} \right)_r \right] \Psi(\mathbf{r}, \mathbf{R}) = 0. \tag{15.4}$$

Here, $h(\mathbf{r})$ is the Hamiltonian of the hydrogen atom, and the interaction
between $\bar{p}$ and H is given by

$$V = -\frac{1}{R} + \frac{1}{|\mathbf{r} - \mathbf{R}|}. \tag{15.5}$$

Departing from a single-center expansion, the wave function is written as

$$\Psi = \sum_i c_i(b, t) \phi_i(\mathbf{r}) \, \mathrm{e}^{-\mathrm{i}\epsilon_i t}, \tag{15.6}$$

where ϕ_i is an atomic wave function with energy ϵ_i, see Eq. (2.39). In order
to take account of the symmetry with respect to the collision plane spanned
by the vectors $\mathbf{b}$ and $\mathbf{v}$, one combines spherical harmonics so that the sum
is symmetric with respect to reflections on the collision plane ($y \to -y$).
Hence

$$\phi_i \equiv \phi_{nlm} = r^{-1} R_{nl}(r) \left[(-1)^m Y_{lm}(\hat{\mathbf{r}}) + Y_{l,-m}(\hat{\mathbf{r}}) \right] \left(2(1 + \delta_{m0}) \right)^{-1/2}. \tag{15.7}$$

The radial functions R_{nl} are expanded [237] into a Laguerre basis of square-
integrable functions [238] as

$$R_{nl}(r) = r^{l+1} \exp(-\alpha r) \sum_k a_{kn}^{(l)} L_k^{2l+2}(2\alpha r), \tag{15.8}$$

where L_k^{2l+2} is an associated Laguere polynomial (2.46) and α is an arbitrary constant. The underlying basis is related to the Sturmian basis
(2.50). The coefficients $a_{kn}^{(l)}$ are obtained by diagonalizing the Hamiltonian
h of atomic hydrogen. The basis functions for a given orbital angular momentum l contain contributions from all bound and continuum states with
the same angular momentum. Owing to the diagonalization procedure,
they are orthogonal and free from overcompleteness. The convergence of
the Laguerre expansion has been studied systematically in [238] and is
found to be essentially independent of the parameter α.

By inserting Eqs. (15.6) and (15.7) into the Schrödinger equation (15.4), we obtain the coupled equations

$$i\,\dot{c}_i(t) = \sum_j c_j(t)\,\langle\phi_i|V|\phi_j\rangle\,e^{i(\epsilon_i - \epsilon_j)t}. \tag{15.9}$$

In contrast to the general case expressed by Eq. (2.8) and the two-center expansion (2.38), the basis states are orthogonal and normalized, so that the overlap matrix reduces to the unit matrix. With the appropriate initial conditions and the choice $\alpha = 0.6$, the cross section for ionization is given by Eqs. (2.5) and (2.6). Four different basis sets have been used, the results of which are in close agreement above 25 keV. With the largest basis, good convergence is achieved above 1 keV, see Fig. 15.1. The agreement with other single-center coupled-channel calculations [239, 240] is good, however, the results of calculations on a three-dimensional lattice [241] are about 10 % higher. The CDW-EIS approach, see Sec. 3.5, is valid only above 100 keV. Excitation cross sections and angle-differential cross sections have also been calculated [237].

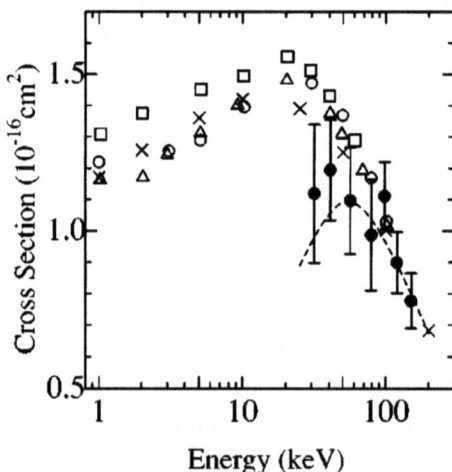

Figure 15.1: Total ionization cross section in $\bar{p} + $ H collisions. Crosses: calculations of [237]; open circles: [239]; triangles: [240]; squares: [241]; dashed line: CDW-EIS model [242]; full circles: experimental data [243]. From [237].

Although antiprotons cannot bind electrons, the introduction of a second center in coupled-channel calculations is expected to modify the results as compared to single-center calculations, because the repulsion by the projectile will distort the wavefunction. It has been shown by Toshima [244], that the addition of continuum states centered at the projectile accelerates the convergence considerably. Owing to the repulsion by the antiproton, the electron density distributions exhibit dips near the antiproton. It is seen from the calculations that below 1 keV antiproton energy the single-center calculations are not converged even for the largest basis sets, and at 0.1 keV, the two-center cross section is larger than the single-center one by a factor of 1.4. Perturbative calculations within the CDW-EIS approximation, see Sec. 3.5, for intermediate velocities have also been performed [245].

15.2 Antiprotonic helium

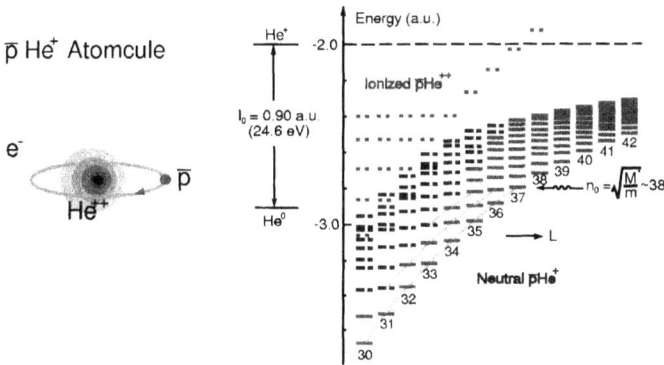

Figure 15.2: (a) Left-hand panel: The structure of $\bar{p}\,\mathrm{He}^+$, in which $\bar{p}$ orbits around the He^{2+} nucleus, while the electron occupies the distributed 1s state. (b) Right-hand panel: The level scheme for large-(n, l) states of $\bar{p}\,\mathrm{He}^+$. The solid bars indicate radiation-dominated metastable states, while the broken bars stand for Auger-dominated short-lived states. The ionized states of $\bar{p}\,\mathrm{He}^+$ are shown by dotted lines. From [247].

A surprising effect emerged when in 1991 it was discovered that antiprotons were able to survive in an environment of helium atoms for time intervals of the order of microseconds [246]. How does this happen? A detailed review on this subject has recently been published by T. Yamazaki et al. [247].

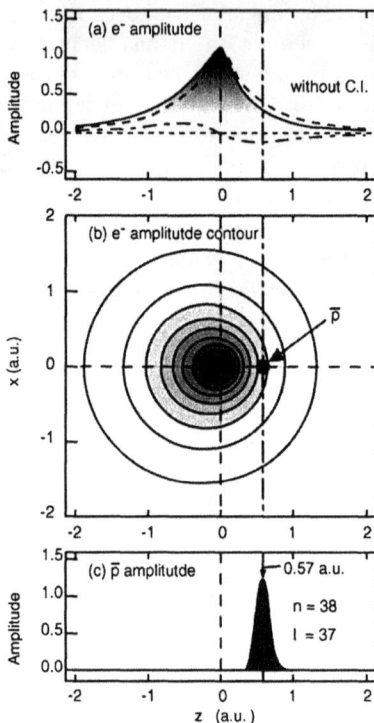

Figure 15.3: Spatial distribution of e^- and $\bar{p}$ in a typical circular state $(n, l) =$ (38, 37). (a) Amplitude of e^-, (b) amplitude contour of e^-, (c) amplitude of $\bar{p}$ located on the z axis. From [247, 248].

The kinetic energy of an antiproton slowing down in a He environment at some stage falls below the He ionization energy of 24.6 eV, at which point it replaces one of the electrons in a He atom. The newly formed $\bar{p}\text{He}^+$ atom recoils with an initial kinetic energy of about 5 eV and continues its journey through the helium medium until, after a time shorter than a nanosecond, it reaches thermal equilibrium without suffering destruction. While the remaining electron is in the 1s ground state, the captured antiproton occupies a highly excited (n, l) state.

In analogy to Eq. (15.3), the principal quantum number is $n \approx \sqrt{\mu_{\bar{p}\text{He}}/m_e} \approx 38$, where $\mu_{\bar{p}\text{He}}$ is the reduced mass of the $\bar{p}$-He system. The orbital angular momentum l brought in by the captured $\bar{p}$ can be as large as $l \approx n - 1$ corresponding to a circular orbit. As illustrated in Fig. 15.2, the $\bar{p}$ orbits the helium nucleus on a well-localized classical orbit, while the 1s electron density is distributed according to quantum mechanics.

While normally, the neutral atom would rapidly ionize by Auger transitions to the continuum, this path is suppressed by the atom's high ionization energy of about 25 eV, in comparison with the $n \rightarrow n-1$ level spacings of about 2 eV. The Auger transitions of sufficient energy from near-circular states ($l \approx n - 1$) are associated with large angular momentum jumps and thus are highly hindered. The main decay mode of circular states is via the cascade $(n, l) \rightarrow (n - 1, l - 1)$ along closely-spaced levels with a typical level lifetime of about 1.5 μs.

The theoretical description of antiprotonic helium may proceed in two possible ways. (1) One considers the system as an atom composed of the He^{2+} nucleus binding an antiproton in a high (n, l) state and an electron in the 1s ground state [248]. The presence of the localized antiproton then gives rise to a polarization of the electron wave function to be described by configuration interactions, mainly by the admixture of 2p states. Conversely, the electron modifies the antiproton orbit. The distribution of electron and antiproton wave functions is illustrated in Fig. 15.3.

(2) An alternative approach takes into account that both the helium nucleus and the antiproton are heavy particles and move slowly compared to the electron. Hence the system behaves more like a diatomic molecule, see Shimamura [234]. The system will perform rotations and vibrations. This view forms the basis of more advanced formulations. Korobov [249] developed a variational method using the molecular-type basis functions. He obtained excellent convergence and predicted transition energies to very high precision. Later, relativistic corrections and QED effects were included in the calculations. Kino et al. [250] developed a variational method employing three types of coupling schemes among the three particles taking into account the correlations on equal footing. Their calculational precision was a high as Korobov's.

Because the transition frequencies between adjacent metastable energy levels of $\bar{p}He^+$ are in the visible range, it is possible to perform precision laser resonance spectroscopy. The basic idea relies on the observation, see Fig. 15.2 (b), that in the vicinity of long-lived states there are always short-lived levels with lifetimes dominated by Auger transitions. These states ionize very rapidly, and the subsequent collision-induced annihilation of the resulting $\bar{p}He^{2+}$ is almost instantaneous and can be detected with 100% efficiency. In this way, it is possible to measure energy differences between long-lived states (solid bars in Fig. 15.2 (b)) and short-lived states (dashed bars) with high precision.

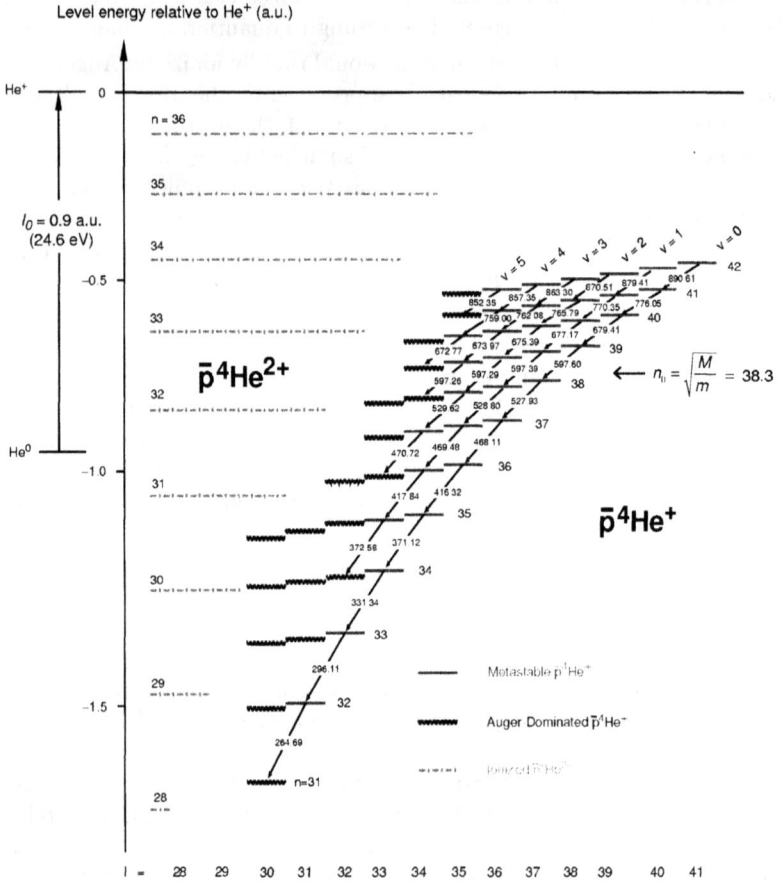

Figure 15.4: Level diagram of $\bar{p}\,^4\mathrm{He}^+$ in relation to $\bar{p}\,^4\mathrm{He}^{2+}$. The solid and wavy bars stand for metastable and short-lived states, respectively, and the dash-dotted lines are are for the l-degenerate ionized states. From [247].

With this experimental accuracy, it turned out that the original nonrelativistic calculations [234, 249] were not accurate enough (showing a discrepancy with experiment for the transition energies on the 1000 ppm level) and that relativistic and QED corrections had to be introduced [251, 252, 253]. The converged results for the energy values have an accuracy of the order of 10^{-14}. With the relativistic and QED corrections, the agreement of the transition energies with experimental data is now of the order of ppm.

We here do not discuss the various side effects that will occur and that are well understood, like Auger lifetimes, collisonal quenching, admixture of foreign molecules etc, see [247] for all the details. We just show in Fig. 15.4 a final level diagram for $\bar{p}\,{}^{4}\mathrm{He}^{+}$ including the transition energies. A similar diagram has also been measured and calculated for $\bar{p}\,{}^{3}\mathrm{He}^{+}$, for which the different effective mass leads to some modifications.

Appendix A

Fundamental constants and units

Physical quantity	Symbol	Numerical value
speed of light in vacuum	c	299 792 458 m s^{-1}
Planck's constant	$\hbar$	1.054 571 68 (18) 10^{-34} J s
		6.582 119 15 (56) 10^{-16} eV s
elementary charge	e	1.602 176 53 (14) 10^{-19} C
electron mass	m_e	9.109 382 6 (16) 10^{-31} kg
electron mass in u	m_e	5.485 799 094 5 (24) u
electron mass energy equivalent	$m_e c^2$	0.510 998 918 (44) MeV
proton mass	M_p	1.672 621 71 (29) 10^{-27} kg
proton mass in u	M_p/u	1.007 276 466 88 (13)
atomic mass unit	u	1.660 538 86 (28) 10^{-27} kg
atomic mass unit in m_e	u/m_e	1822. 888 5
atomic mass unit in MeV/c^2	$u/(\text{MeV}/c^2)$	931.494 043 (80)
electron Compton wave length	$\lambdabar_c$	3.861 592 678 (26) 10^{-13} m
proton Compton wave length	$\lambdabar_{cp}$	2.103 089 104 (14) 10^{-16} m
classical electron radius	r_0	2.817 940 325 (28) 10^{-15} m
Bohr radius	a_0	5.291 772 108 (18) 10^{-11} m
inverse fine-structure constant	$1/\alpha$	137.035 999 11 (46)
Hartree energy	E_h	27.211 384 5 (23) eV

From NIST Reference on Constants, Units and Uncertainty, 2005. The numbers in parentheses denote the uncertainties in the last digits.

Atomic units: $\hbar = 1$, $m_e = 1$, $e = 1$

Physical quantity	Symbol	Numerical value
unit of mass	m_e	$9.109\ 382\ 6\ 10^{-31}$ kg
unit of length	$a_0 = \hbar^2/m_e e^2$	$5.291\ 772\ 108\ 10^4$ fm
unit of area	a_0^2	$2.800\ 285\ 10^7$ b
unit of time	$\hbar^3/m_e e^4$	$2.418\ 884\ 326\ 505\ 10^{-17}$ s
unit of velocity	$e^2/\hbar = \alpha c$	$2.187\ 691\ 263\ 3\ 10^6$ m s^{-1}
unit of energy	$e^2/a_0 = m_e e^4/\hbar^2$	$27.211\ 384\ 5$ eV
unit of electric charge	e	$1.602\ 176\ 53\ 10^{-19}$ C
unit of electric potential	e/a_0	$27.211\ 384\ 5$ V
unit of electric field	e/a_0^2	$5.142\ 206\ 42\ 10^{-4}$ V/fm

Natural (relativistic) units: $\hbar = 1$, $m_e = 1$, $c = 1$

Physical quantity	Symbol	Numerical value
unit of mass	m_e	$9.109\ 382\ 6\ 10^{-31}$ kg
unit of length	$\lambda_c = \hbar/m_e c$	$386.159\ 267\ 8$ fm
unit of area	λ_c^2	$1.491\ 189\ 10^3$ b
unit of time	$\lambda_c/c = \hbar/m_e c^2$	$1.288\ 088\ 667\ 7\ 10^{-21}$ s
unit of velocity	c	$2.997\ 924\ 58\ 10^8$ m s^{-1}
unit of energy	$m_e c^2$	$0.510\ 998\ 918$ MeV
unit of electric charge	e	$1.602\ 176\ 53\ 10^{-19}$ C
unit of electric potential	$e/\lambda_c = \alpha\, m_e c^2/e$	$3.728\ 940\ 6\ 10^3$ V
unit of electric field	e/λ_c^2	$9.656\ 482\ 0$ V/fm

Bibliography

[1]   M.R.C. Mcdowell and J.P. Coleman, *Introduction to the Theory of Ion-Atom Collisions* (North Holland, Amsterdam, 1970)

[2]   B.H. Bransden and M.R. C. McDowell, *Charge Exchange and the Theory of Ion-Atom Collisions* (Clarendon Press, Oxford, 1992)

[3]   J. Eichler and W.E. Meyerhof, *Relativistic Atomic Collisions*, (Academic Press, San Diego, 1995).

[4]   H. Winter, Phys. Rep. **367** (2002) 387.

[5]   R. Dörner, V. Mergel, O. Jagutzki, L. Spielberger, J. Ullrich, R. Moshammer, and H. Schmidt-Böcking, Phys. Rep. **330** (2000) 96.

[6]   J. Ullrich, R. Moshammer, A. Dorn, R. Dörner, L.Ph.H. Schmidt, H. Schmidt-Böcking, Reports on Progress in Physics, **66** (2003) 1463.

[7]   R. Abrines and I.C. Percival, Proc. Phys. Soc. **88** (1966) 861, 873.

[8]   R.E. Olson and A. Salop, Phys. Rev. A **16** (1977) 531.

[9]   H. Goldstein, *Classical Mechanics*, 2nd ed. (Addison Wesley, Reading 1981).

[10]   V. Fock, Z. Phys. **98** (1935) 145.

[11]   W. Fritsch and C.D. Lin, Phys. Rep. **202** (1991) 1.

[12]   D.R. Bates and R. McCarroll, Prog. Roy. Soc. London A **245** (1958) 175.

[13]   M. Born and R. Oppenheimer, Ann. Physik **84** (1927) 457.

[14]   U. Fano and W. Lichten, Phys. Rev. Lett. **14** (1965) 627.

[15]   J. Eichler, U. Wille, B. Fastrup, and K. Taulbjerg, Phys. Rev. A **14** (1976) 707.

[16]　　J. Eichler and U. Wille, Phys. Rev. Lett. **33** (1974) 56.

[17]　　J. Eichler and U. Wille, Phys. Rev. A **11** (1975) 1973.

[18]　　Q.C. Kessel and B. Fastrup, Case Stud. At. Phys. **3** (1973) 137.

[19]　　F.P. Larkins, J. Phys. B: At. Mol. Phys. **5** (1972) 571.

[20]　　U. Wille and R. Hippler, Phys. Rep. **132** (1986) 129.

[21]　　M. Barat and W. Lichten, Phys. Rev. A. **6** (1972) 211.

[22]　　B. Fricke, T. Morovic, W.-D. Sepp, A. Rosen and D.E. Ellis, Phys. Lett. **59A** (1976) 375.

[23]　　W.-D. Sepp, B. Fricke and T. Morovic, Phys. Lett. **81A** (1981) 258.

[24]　　B. Fricke and W.-D. Sepp, J. Phys. B **14** (1981) L549.

[25]　　R. Schuch, M. Meron, B.M. Johnson, K.W. Jones, R. Hoffmann, H. Schmidt-Böcking and I. Tserruya Phys. Rev. A **37** (1988) 3313.

[26]　　H. Nakamura and C. Zhu, Comments At. Mol. Phys. **32** (1996) 249.

[27]　　E.C.G. Stückelberg, Helv. Phys. Acta **5** (1932) 369.

[28]　　H. Danared and A. Bárány, J. Phys. B **19** (1986) 3109.

[29]　　M. Barat, P. Roncin, L. Guillemot, M.N. Gaboriaud, and H. Laurent, J. Phys. B **23** (1990) 2811.

[30]　　N. Keller, L.R. Andersson, R.D. Miller, M. Westerlind, S.B. Elston, I.A. Sellin, C. Biedermann, and H. Cederquist, Phys. Rev. A **48** (1993) 3684.

[31]　　M. Hoshino, Y. Yamazaki et al. private communication, 2004.

[32]　　A. Bárány, H. Danared, H. Cederquist, P. Hvelplund, H. Knudsen, J.O.K. Pedersen, C.L. Cocke, L.N. Tunnell, W. Waggoner and J.P. Giese,J. Phys. B **19** (1986) L427.

[33]　　M. Hoshino, Y. Kanai, F. Mallet, Y. Nakai, M. Kitajima, H. Tanaka and Y. Yamazaki, Nucl. Instr. Meth. Phys. Res. B **205** (2003) 568.

[34]　　T.G. Winter and C.D. Lin, Phys. Rev. A **29** (1984) 567.

[35]　　W.Fritsch and C.D. Lin, Phys. Rev. A **82** (1982) 762.

[36]　　W. Fritsch and C.D. Lin, Phys. Rev. A **27** (1983) 3361.

[37]　　E. Merzbacher, *Quantum Mechanics 3rd ed* (Wiley, New York, 1998)

[38]   I.S. Gradshteyn and I.M. Ryzhik, *Table of Integrals, Series, and Products*, (Academic Press, New York, 1980)

[39]   H.A. Bethe and E.E. Salpeter, *Quantum Mechanics of One- and Two-Electron Atoms* (Plenum, New York, 1977).

[40]   M. Rotenberg, Ann. Phys. **19** (1962) 262.

[41]   D.F. Gallaher and L. Wilets, Phys. Rev. **169** (1968) 139.

[42]   N. Toshima and J. Eichler, Phys. Rev. Lett. **66** (1991) 1050., Phys. Rev. A **46** (1992) 2564.

[43]   K. Gramlich, N. Grün, and W. Scheid, J. Phys. B **22** (1989) 2567.

[44]   F.J. Dyson, Phys. Rev. **75** (1949) 486.

[45]   W. Magnus, Comm. Proc. Appl. Math. **7** (1954) 649.

[46]   P. Pechukas and J.C. Light, J. Chem. Phys. **44** (1966) 3897.

[47]   K. Alder and A. Winter, Mat. Fys. Medd. Dan. Vid. Selsk. **32**, no.8 (1960)

[48]   J. Eichler, Phys. Rev. A **15** (1977) 1856.

[49]   D.P. Dewangan and J. Eichler, Phys. Rep. **247** (1994) 59.

[50]   B. Crasemann, ed. *Atomic Inner-Shell Processes, Vol. I: Ionization and Transition Probabilities*, (Academic Press, 1975).

[51]   A. Messiah, *Quantum Mechanics 3rd ed* (Wiley, New York, 1998).

[52]   D. Belkić, R. Gayet, and A. Salin, Phys. Rep. **56** (1979) 279.

[53]   F. Decker and J. Eichler, J. Phys. B **22** (1989) 3023.

[54]   D.S.F. Crothers and J.F. McCann, J. Phys. B **16** (1983) 3229.

[55]   P.D. Fainstein, V.H. Ponce, and R.D. Rivarola, J. Phys. B **27** (1991) 3091.

[56]   L. Gulyàs, P.D. Fainstein, and A. Salin, J. Phys. B **28** (1995) 245.

[57]   E. Clementi and C. Roetti, At. Data Nucl. Data Tables **14** (1974) 177.

[58]   M.E. Rudd, L.H. Toburen, and N. Stolterfoht, At. Data, Nucl. Data Tables **23** (1979) 965.

[59]   D.H. Lee, P. Richard, T.J.M. Zouros, J.M. Sanders, J.M. Shinpaugh and H. Hidmi, Phys. Rev. A **24** (1990) 97.

[60]   N. Toshima, Phys. Rev. A **59** (1999) 1981.

[61]   J.R. Oppenheimer, Phys. Rev. **31** (1928) 349.

[62]   H.C. Brinkman and H.A. Kramers, Proc. Acad. Sci. Amsterdam **33** (1930) 973.

[63]   J.D. Jackson and H. Schiff, Phys. Rev. **89** (1953) 359.

[64]   J.S. Briggs, J. Macek and K. Taulbjerg, Comments At. Mol. Phys. **12** (1982) 1.

[65]   J. Macek and S. Alston, Phys. Rev. A **26** (1982) 250.

[66]   J.H. Macek and R. Shakeshaft, Phys. Rev. A **22** (1980) 1441.

[67]   J.H. Macek and K. Taulbjerg, Phys. Rev. Lett. **46** (1981) 170.

[68]   J.S. Briggs, J. Phy. B **10** (1977) 3075.

[69]   D.P. Dewangan and J. Eichler, J. Phys. B **18** (1985) L65.

[70]   D.P. Dewangan and J. Eichler, Comments At. Mol. Phys. **21** (1987) 1.

[71]   D.P. Dewangan and J. Eichler, J. Phys. B **19** (1986) 2939.

[72]   D.P. Dewangan and J. Eichler, Comments. At. Mol Phys. **21** (1987) 1. Nucl. Instr. Meth. Phys. Res. **B23** (1987) 160.

[73]   L.J. Dubé, B. Mensour, D.P. Dewangan and H.S. Chakraborty, J. Phys. B **23** (1990) L711.

[74]   F. Decker and J. Eichler, Phys. Rev. A **39** (1989) 1530.

[75]   N. Toshima, T. Ishihara, and J. Eichler, Phys. Rev.A **36** (1987) 2659.

[76]   I.M. Cheshire, Proc. Phys. Soc. **84** (1964) 89.

[77]   F.T. Chan and J. Eichler, Phys. Rev. Lett. **42** (1979) 58.

[78]   J. Eichler and F.T. Chan, Phys. Rev. A **20** (1979) 104.

[79]   F.T. Chan and J. Eichler, Phys. Rev. A **20** (1979) 1841.

[80]   J.K.M. Eichler, Phys. Rev. A **23** (1981) 498.

[81]   N. Toshima, J. Phys. B **32** (1999) L615.

[82]   L.H. Thomas, Proc. Roy. Soc. (London) **A114** (1927) 561.

[83]   R. Shakeshaft and L. Spruch, Rev. Mod. Phys. **51** (1979) 369.

[84]   R.M. Drisko, Ph.D. Thesis, Carnegie Institute of Technology, 1955.

[85]   Dz. Belkić, Europhys. Lett. **7** (1988) 323.; Phys. Rev. A **43** (1991) 4751.

[86]   H. Vogt, R. Schuch, T.E. Justiniano, M. Schutz, and W. Schwab, Phys. Rev. Lett. **57** (1986) 2256.

[87]   J.D. Jackson, *Classical Electrodynamics*, 2nd ed. (Wiley, New York, 1975)

[88]   J.D. Bjorken and S.D. Drell, *Relativistic Quantum Mechanics* (McGraw-Hill, New York, 1964).

[89]   V.B. Berestetskii, E.M. Lifshitz, and L.P. Pitaevskii, *Quantum Electrodynamics* (Pergamon, Oxford, 1982).

[90]   J.J. Sakurai, *Advanced Quantum Mechanics* (Addison Wesley, Reading, 1967).

[91]   K.G. Dedrick, Revs. Mod. Phys. **34** (1962) 429.

[92]   R.G. Newton, *Scattering Theory of Waves and Particles* (McGraw-Hill, New York, 1966).

[93]   D.C. Ionescu, private communication 2003 as a 3-dimensional adaption of a graph in Ref. [3].

[94]   E. Fermi, Physik **29** (1924) 315.

[95]   C.F. v. Weizsäcker, Z. Physik **88** (1934) 612.

[96]   E.J. Williams, Phys. Rev. **45** (1934) 729. Dan. Vid. Selsk. Mat.-Fys. Medd. **13** (1935) No. 4.

[97]   C.A. Bertulani and G. Baur, Phys. Rep. **163** (1988) 299.

[98]   G. Baur, K. Hencken, D. Trautmann, S, Sadovsky, and Yu. Kharlov, Phys. Rep. **364** (2002) 359.

[99]   M.E. Rose, *Relativistic Electron Theory* (Wiley, New York, 1961).

[100]   M.E. Rose, *Elementary Theory of Angular Momentum* (Wiley, New York, 1957).

[101]   C.G. Darwin, Proc. Roy. Soc. London, Ser. A **118** (1928) 654.

[102]   A. Sommerfeld and A.W. Maue, Ann. Physik **22** (1935) 629.

[103]   W.H. Furry, Phys. Rev. **46** (1934) 391.

[104]   H.A. Bethe and L. Maximon, Phys. Rev. **93** (1954) 768.

[105]   D.C. Ionescu, private communication 2003.

[106]   N. Toshima and J. Eichler, Phys. Rev. Lett. **60** (1988) 573.

[107]   N. Toshima and J. Eichler, Phys. Rev. A **38** (1988) 2305.

[108]   J. Eichler, Phys. Rev. A **35**, 3248 (1987); Erratum: **37** (1988) 287.

[109]   N. Toshima and J. Eichler, Phys. Rev. A **42** (1990) 3896.

[110]   U. Becker, Ph.D. Thesis, Giessen (1986), see in *Physics of Strong Fields*, ed. W. Greiner, NATO Advanced Study Institute Series B: Physics, Vol. 153, p. 609 (Plenum, New York, 1987).

[111]   J. Thiel, A. Bunker, K. Momberger, N. Grün, and W. Scheid, Phys. Rev. A **46** (1992) 2607.

[112]   O. Busic, N. Grün and W. Scheid, Phys. Rev. A **70** (2004) 062707.

[113]   K. Momberger, A. Belkacem, and A.H. Sørensen, Phys. Rev. A **53** (1996) 1605.

[114]   D.C. Ionescu, Habilitationsschrift, Freie Univesität Berlin, 1997.

[115]   D.C. Ionescu and A. Belkacem, Eur. Phys. J. D, **18** (2002) 301.

[116]   P.A. Amundsen and K. Aashamar, J. Phys. B **14** (1981) 4047.

[117]   B.L. Moiseiwitsch, Phys. Reports, **118** (1985) 135.

[118]   Th. Stöhlker, D.C. Ionescu, P. Rymuza, T. Ludziejewski, P.H. Mokler, H. Geissel, C. Scheidenberger, F. Bosch, B. Franzke, O. Klepper, C. Kozhuharov, R. Moshammer, F. Nickel, H. Reich, Z. Stachura, A. Warczak, Nucl. Instr. Meth. Phys. Res. B **124** (1997) 160.

[119]   R. Anholt, Phys. Rev. A **19** (1979) 1004.

[120]   U. Becker, N. Grün and W. Scheid, J. Phys. B **18** (1985) 4589.

[121]   R. Anholt and U. Becker, Phys. Rev. A **36** (1987) 4628.

[122]   J. Eichler, Phys. Rep. **193** (1990) 165.

[123]   A.B. Voitkiv, Phys. Rep. **392** (2004) 191.

[124]   Th. Stöhlker, D. C. Ionescu, P. Rymuza, F. Bosch, H. Geissel, C. Kozhuharov, T. Ludziejewski, P.H. Mokler, C. Scheidenberger, Z. Stachura, A. Warczak, R.W. Dunford, Physics Lett. A **238** (1998) 43., Phys. Rev. A **57** (1998) 845.

[125]   R. Shakeshaft, Phys. Rev. A **20** (1979) 779.

[126]   B.L. Moiseiwitsch and S.G. Stockman, J. Phys. B **13** (1980) 2975.

[127]   F. Decker and J. Eichler, Phys. Rev. A **44** (1991) 377.

[128]   N. Toshima and J. Eichler, Phys. Rev. A **41** (1990) 5221.

[129]   N. Toshima and J. Eichler, Comments At. Mol. Phys. **31** (1995) 109.

[130]   J. Eichler, Phys. Rev. A **32** (1985) 112.

[131]   A. Ichihara, T. Shirai, and J. Eichler, At. Data and Nucl. Data Tables. **55** (1993) 63.

[132]   W.E. Meyerhof, R. Anholt, J. Eichler, H. Gould, Ch. Munger, J. Alonso, P. Thieberger, and H.E. Wegner, Phys. Rev. A **32** (1985) 3291.

[133]   A. Belkacem, H. Gould, B. Feinberg, R. Bossingham and W.E. Meyerhof, Phys. Rev. A **56** (1997) 2806.

[134]   T. Brunne, Doctoral Thesis, Freie Universität Berlin, 2001.

[135]   J. Eichler and Th. Stöhlker, Review in preparation for Physics Reports

[136]   M. Kleber and D.H. Jakubassa, Nucl. Phys. A **252** (1975) 152.

[137]   J. Eichler, A. Ichihara, and T. Shirai, Phys. Rev.A **51** (1995) 3027.

[138]   W. Heitler, *The Quantum Theory of Radiation* (Oxford University Press, Oxford, 1954).

[139]   E. Spindler, H.-D. Betz, and F. Bell, Phys, Rev. Lett. **42** (1979) 832. E. Spindler, Ph.D. Thesis, München, 1979.

[140]   R.H. Pratt, A. Ron, and H.K. Tseng, Revs. Mod. Phys. **45** (1973) 273.

[141]   M. Stobbe, Ann. Phys. (Leipzig) **7** (1930) 661.

[142]   A. Burgess, Mem. Roy. Ast. Soc. **69** (1964) 1.

[143]   A. Ichihara, T. Shirai, and J. Eichler, Phys. Rev. A **49** (1994) 1875.

[144]   J. Eichler, A. Ichihara, and T. Shirai, Phys. Rev A **58** (1998) 2128.

[145]   A. Ichihara, private communication of data 1998.

[146]   A. Ichihara, private communication 2002.

[147]   A.E. Klasnikov, A.N. Artemyev, T. Beier, J. Eichler, V.M. Shabaev, and V.A. Yerokhin, Phys. Rev. A **66** (2002) 042711.

[148]   Th. Stöhlker, T. Ludziejewski, F. Bosch, R.W. Dunford, C. Kozhuharov, P.H. Mokler, H.F. Beyer, O.Brinzanescu, F. Franzke, J. Eichler, A. Griegal, S. Hagmann, A. Ichihara, A. Krämer, D. Liesen, H. Reich, P. Rymuza, Z. Stachura, M. Steck, P. Swiat, and A. Warczak, Phys. Rev. Lett. **82**, 3232 (1999).

[149]   Th. Stöhlker, H. Geissel, H. Irnich, T. Kandler, C. Kozhuharov, P.H. Mokler, Münzenberg, F. Nickel, C. Scheidenberger, T. Suzuki, M. Kucharski, A. Warczak, P. Rymuza, Z. Stachura, A. Kriessbach, D. Dauvergne, B. Dunford, J. Eichler, A. Ichichara, T. Shirai, Phys. Rev. Lett. **73** (1994) 3520.

[150]   Th. Stöhlker, X. Ma, T. Ludziejewski, H.F. Beyer, F. Bosch, O. Brinzanescu, R.W. Dunford, J. Eichler, S. Hagmann, A. Ichihara, C. Kozhuarov, A. Krämer, D. Liesen, P.H. Mokler, Z. Stachura, P. Swiat, and A. Warczak, Phys. Rev. Lett. **86** (2001) 983.

[151]   A. Ichihara, T. Shirai and J. Eichler, Phys. Rev. A **54** (1996) 4954. At the highest energies, 10 and 10.8 GeV/u, the results were not fully converged in the last digit.

[152]   Th. Stöhlker, F. Bosch, A. Gallus, C. Kozhuharov, G. Menzel, P.H. Mokler, H.T. Prinz, J. Eichler, A. Ichihara, T. Shirai, R.W. Dunford, T. Ludziejewski, P. Rymuza, Z. Stachura, P. Swiat, and A. Warczak, Phys. Rev. Lett. **79** (1997) 3270.

[153]   A. Surzhykov, S. Fritzsche, A. Gumberidze, and Th. Stöhlker, Phys. Rev. Lett. **88** (2002) 153001.

[154]   J. Eichler and A. Ichihara, Phys. Rev. A, **65** (2002) 052716.

[155]   A.N. Artemyev, T. Beier, J. Eichler, A.E. Klasnikov, C. Kozhuharov, V.M. Shabaev, T. Stöhlker, and V.A. Yerokhin, Phys. Rev. A **67** (2003) 052711.

[156]   P. Zimmerer, N. Grün, and W. Scheid, Phys. Lett. A **148**, 457 (1990).

[157]   N. R. Badnell and M. S. Pindzola, Phys. Rev. A **45**, 2820 (1992).

[158]   K. Hencken, private communication, 2002.

[159]   C. Bottcher and M.R. Strayer, Phys. Rev. D **39** (1989) 1330.

[160]   J.C. Wells, V.E. Oberacker, A.S. Umar, C. Bottcher, M.R. Strayer, J.-S. Wu, and G. Plunien, Phys. Rev. A **45** (1992) 6296.

[161]   K. Hencken, D. Trautmann, and G. Baur, Phys. Rev. A **51** (1995) 998., **51** (1995) 1874.

[162]   A. Aste, G. Baur, K. Hencken, D. Trautmann and G. Scharf, Eur. Phys. J. C **23** (2002) 545.

[163]   G. Racah, Nuovo Cimento, **14** (1937) 93.

[164]   M.C. Güçlü, J. Li, A.S. Umar, D.J. Ernst, and M.R. Strayer, Annals of Physics **272** (1999) 7.

[165]   U. Becker, N. Grün, and W. Scheid, J. Phys. B **19** (1986) 1347.

[166]   F. Decker, Phys. Rev. A **44** (1991) 2883.

[167]   D.C. Ionescu and J. Eichler, Phys. Rev. A **48** (1993) 1176.

[168]   C.R. Vane, S. Datz, P.F. Dittner, H.F. Krause, C. Bottcher, M. Strayer, R. Schuch, H. Gao, and R. Hutton, Phys. Rev. Lett. **69** (1992) 1911.

[169]   P.B. Eby, Phys. Rev. A **43** (1991) 2258.

[170]   A. Belkacem, H. Gould, B. Feinberg, R. Bossingham, and W.E. Meyerhof, Phys. Rev. Lett. **72** (1993) 1514.

[171]   A. Belkacem, H. Gould, B. Feinberg, R. Bossigham, and W.E. Meyerhof, Phys. Rev. Lett. **73** (1994) 2432.

[172]   C.R. Vane, S. Datz, E.F. Deveney, P.F. Dittner, H.F. Krause, R. Schuch, H. Gao, and R. Hutton, Phys. Rev. A **56** (1997) 3682.

[173]   H. Gould, in *Physics of Electronic and Atomic Collisions*, Proceedings of the XX International Conference on the Physics of Electronic and Atomic Collisions, Vienna 1997.

[174]   U. Becker, N. Grün, and W. Scheid, J. Phys. B **20** (1987) 2075.

[175]   U. Becker, J. Phys. B **20** (1987) 6563.

[176]   A.J. Baltz, M.J. Rhoades-Brown, and J. Weneser, Phys. Rev. A **44** (1991) 5569.

[177]   H. Meier, Z. Halabuka, K, Hencken and D. Trautmann, Phys. Rev. A **63** (2001) 032713.

[178]   H.J. Bär and G. Soff, Physica **128C** (1985) 225.

[179]   K. Rumrich, K. Momberger, G. Soff, W. Greiner, N. Grün, and W. Scheid, Phys. Rev. Lett. **66** (1991) 2613.

[180]   K. Rumrich, G. Soff, and W. Greiner, Phys. Rev. A **47** (1993) 215.

[181]   M. Gail, N. Grün and W. Scheid, J. Phys. B **36** (2003) 1397.

[182]   J.C. Wells, V.E. Oberacker, M.R. Strayer, and A.S. Umar, Phys. Rev. A **53** (1996) 1498.

[183]   D.C. Ionescu, Phys. Rev. A **49** (1994) 3188.

[184]   J. Eichler, Phys. Rev. Lett. **75** (1995) 3653.

[185]   D.C. Ionescu and J. Eichler, Phys. Rev. A **54** (1996) 4960.

[186]   A.J. Baltz, M.J. Rhoades-Brown, and J. Weneser, Phys. Rev. A **47** (1993) 3444.

[187]   A.J. Baltz, M.J. Rhoades-Brown, and J. Weneser, Phys. Rev. A **50** (1994) 4842.

[188]   A.J. Baltz, Phys. Rev. A **52** (1995) 4970.

[189]   A.J. Baltz, M.J. Rhoades-Brown, and J. Weneser, Phys. Rev. E **54** (1996) 2433.

[190]   A.J. Baltz, Phys. Rev. Lett. **78** (1997) 1231.

[191]   B. Segev and J.C. Wells, Phys. Rev. A **57** (1998) 1849.

[192]   J. Eichler and A. Belkacem, Phys. Rev. A **54** (1996) 5427.

[193]   J.C. Wells, B. Segev, and J. Eichler, Phys. Rev. A **59** (1999) 346.

[194]   H. Davies, H.A. Bethe, and L.C. Maximon, Phys. Rev. **93** (1954) 788.

[195]   G. Baur, Phys. Rev. A **42** (1990) 5736.

[196]   M.C. Güçlü, J.C. Wells, A.S. Umar, M.R. Strayer, and D.J. Ernst, Phys. Rev. A **51** (1995) 1836.

[197]   A. Alscher, K. Hencken, D. Trautmann, and G. Baur, Phys. Rev. A **55** (1997) 396.

[198]   M.J. Rhoades-Brown and J. Weneser, Phys. Rev. A **44** (1991) 330.

[199]   Ch. Best, W. Greiner, and G. Soff, Phys. Rev. A **46** (1992) 261.

[200]   U. Fano, D. Green, J.L. Bohn and T.A. Heim, J. Phys. B **32** (1999) R1.

[201]   C.D. Lin, Phys. Rep. **257** (1995) 1.

[202] P.M. Morse and H. Feshbach, *Methods of Theoretical Physics* (McGraw-Hill, New York 1953), p. 1754.

[203] J. Macek, J. Phys. B **1** (1968) 831.

[204] C-N. Liu, A-T. Le, T. Morishita, B.D. Esry, and C.D. Lin, Phys. Rev. A **67** (2003) 052705.

[205] A. Igarashi, M. Kimura, and I Shimamura, Phys. Rev. Lett. **89** (2002) 123201.

[206] A. Igarashi and I. Shimamura, Phys. Rev. A **58** (1998) 1166.

[207] K. Kobayashi, T. Ishihara, and N. Toshima, Muon Catal. Fusion **2** (1988) 191.

[208] O.I. Tolstikhin and C. Namba, Phys. Rev. A **60** (1999) 5111.

[209] O. Sørensen, D.V. Fedorov, A.S. Jensen, and E. Nielsen, Phys. Rev. A **65** (2002) 051601.

[210] B.D. Esry and H.R. Sadeghpour, Phys. Rev. A **03** (2003) 012704.

[211] S. Ninomiya, Y. Yamazaki, F. Koike, H. Masuda, T. Azuma, K. Komaki, K. Kuroki, and M. Sekiguchi, Phys. Rev. Lett. **78** (1997) 4557.

[212] R. Morgenstern, private communication 2004.

[213] Y. Yamazaki, Nucl. Instrum. Methods Phys. Res. B **193** (2002) 516.

[214] H. Masuda and K. Fukuda, Science **268** (1995) 1466.

[215] J. Burgdörfer, P. Lerner, and F.W. Meyer, Phys. Rev. A **44** (1991) 5674.

[216] J. Burgdörfer, in C.D. Lin (Ed),*Review of Fundamental Processes and Applications of Atoms and Ions*, (World Scientific, Singapore, 1993) p.517.

[217] K. Tokési, L. Wirtz, C. Lemell, and J. Burgdörfer, Phys. Rev. A **64** (2001) 042902.

[218] Y. Yamazaki and K. Kuroki, Current Opinion in Solid State and Material Science, **6** (2002) 169.

[219] H. Masuda, H. Yamada, M. Satoh, , H. Asoh, M. Nakao, and T. Tamamura, Appl. Phys. Lett. **71** (1997) 2770.

[220] Y. Yamazaki, S. Ninomiya, F. Koike, H. Masuda, T. Azuma, K. Komaki, and M. Sekiguchi, J. Phys, Soc. Jpn. **65** (1996) 1199.

[221]   Y. Yamazaki, Int. J. Mass. Spec. **192** (1999) 437.

[222]   R. Spohr, in *Ion Tracks and Microtechnology*, ed. K. Bethge (Viehweg, Braunschweig, 1990)

[223]   R.L. Fleischer, P.R. Price, and R.M. Walker, *Nuclear Tracks in Solids* (University of California Press, Berkeley, 1975).

[224]   N. Stolterfoht, J.-H. Bremer, V. Hoffmann, R. Hellhammer, D. Fink, A. Petrov, and B. Sulik, Phys. Rev. Lett. **88** (2002) 133201.

[225]   V.V. Okorokov, Yad. Fiz. **2** (1965) 1009., [Sov. J. Nucl. Phys. **2** (1966) 719.]

[226]   S. Datz, C.D. Moak, O.H. Crawford, H.F. Krause, J. Gomez del Campo, J.A. Biggerstaff, P.D. Miller, P. Hvelplund, and H. Knudsen, Phys. Rev. Lett. **40** (1978) 843.

[227]   H.F. Krause and S. Datz, *Channeling Heavy Ions through Crystalline Lattices*, Adv. Atom. Mol. Opt. Phys. **37** (1996) 139.

[228]   K. Komaki, T. Azuma, T. Ito, Y. Takabayashi, Y. Yamazaki, M. Sano, M. Torikoshi, A. Kitagawa, E. Takada, and T. Murakami, Nucl. Instr. Meth. in Phys. Res. B **146** (1998) 19.

[229]   T. Ito, T. Azuma, K. Komaki, and Y. Yamazaki, Nucl. Instr. Meth. Phys. Res. B **135** (1998) 132.

[230]   T. Azuma, T. Ito, K. Komaki, Y. Yamazaki, M. Sano, M. Torikoshi, A. Kitagawa, E. Takada, and T. Murakami, Nucl. Instr. Meth. in Phys. Res. B **135** (1998) 61.

[231]   T. Azuma, T. Ito, Y. Yamazaki, K. Komaki, M. Sano, M. Torikoshi, A. Kitagawa, E. Takada, and T. Murakami, Phys. Rev. Lett. **83** (1999) 528.

[232]   T. Azuma, T. Muranaka, Y. Takabayashi, T. Ito, C. Kondo, K. Komaki, Y. Yamazaki, S. Datz, E, Takada, and T. Murakami, Nucl. Instr. Meth. Phys. Res. B **205** (2003) 779.

[233]   M.H. Holzscheiter, M. Charlton and M.M. Nieto, Phys. Rep. **402** (2004) 1.

[234]   I. Shimamura, Phys. Rev. A **46** (1992) 3776.

[235]   O.I. Tolstikhin, S. Watanabe, and M. Matsuzawa, Phys. rev. A **54** (1996) R3705.

[236]   K. Sakimoto, J. Phys. B **34** (2001) 1769. Phys. Rev. A **65** (2001) 012706. Phys. Rev. A **66** (2002) 032506.

[237]   A. Igarashi, S. Nakazaki, and A. Ohsaki, Phys. Rev. A **61** (2000) 062712.

[238]   I. Bray and A.T. Stelbovics, Phys. Rev. A **46** (1992) 6995.

[239]   K,A. Hall, J.F. Reading, and A.L. Ford, J. Phys. B **29** (1996) 6123.

[240]   G. Schiwietz, U. Wille, R. Muiño Díez, P.D. Fainstein, and P.L. Grande, J. Phys. B **29** (1996) 307.

[241]   J.C. Wells, D.R. Schultz, P. Graves, and M.S. Pindzola, Pgys. Rev. A **54** (1996) 593.

[242]   D.R. Schultz, P.S. Krstić, C.O. Reinhold, and J.C. Wells, Phys. Rev. Lett. **76** (1996) 2882.

[243]   H. Knudsen, U. Mikkelsen, K. Paludan, K, Kirsebom, S.P. Møller, E. Uggerhøi, J. Slevin, M. Carlton, and E. Morenzoni, Phys. Rev. Lett. **74** (1995) 4627.

[244]   N. Toshima, Phys. Rev. A **64** (2001) 024701.

[245]   A.B. Voitkiv and J. Ullrich, Phys. Rev. A **67** (2003) 062703.

[246]   M. Iwazaki, S.N. Nakamura, K. Shigaki, Y. Shimizu, H. Tamura, T. Ishikawa, R.S. Hayano, E. Takada, E. Widmann, H. Outa, M. Aoki, P. Kitching, and T. Yamazaki, Phys. Rev. Lett. **67** (1991) 1246.

[247]   T. Yamazaki, N. Morita, R.S. Hayano, E. Widmann, and J. Eades, Phys. Rep. **366** (2002) 183.

[248]   T. Yamazaki and K. Ohtsuki, Phys. Rev. A **45** (1992) 7782.

[249]   V.I. Korobov, Phys. Rev. A **54** (1996) R1749.

[250]   Y. Kino, M. Kamimura, and H. Kudo, Nucl. Phys. A **631** (1998) 649c.

[251]   V.I. Korobov and D.D. Bakalov, Phys. Rev. Lett. **79** (1997) 3379.

[252]   Y. Kino, M. Kamimura, and H. Kudo, Hyperfine Interactions **119** (1999) 201.

[253]   V.I. Korobov, Nucl. Phys A **684** (2001) 663c. **689** (2001) 75c.

Index

www.ingramcontent.com/pod-product-compliance
Lightning Source LLC
Chambersburg PA
CBHW060350220326
41598CB00023B/2873